Mechanisms of Plant Fertilization

Mechanisms of Plant Fertilization

Editor
Ramniwas Sharma

BIOTECH
2006
Biotech Books
Delhi - 110 035

ISBN 81-7622-160-0

Published by : **BIOTECH BOOKS**
1123/74, Tri Nagar,
DELHI – 110 035
Phone: 27382765
e-mail: biotechbooks@yahoo.co.in

Showroom : 4762-63/23, Ansari Road, Darya Ganj,
NEW DELHI - 110 002
Phone: 23245578, 23244987

Printed at : Tarun Offset Printers,

PREFACE

Plants are generally distinguished from animals in that they possess chlorophyll, are usually fixed in one place, have no nervous system or sensory organs and hence respond slowly to stimuli, and have rigid supporting cell walls containing cellulose. In addition, plants grow continually throughout life and have no maximum size or characteristic form in the adult, as do animals. In higher plants the meristem tissues in the root and stem tips, in the buds, and in the cambium are areas of active growth. Plants also differ from animals in the internal structure of the cell and in certain details of reproduction

In these the gametophyte stage is completely reduced, and the sporophyte begins life inside an enclosure called a seed, which develops while on the parent plant, and with fertilisation by means of pollen grains. Whereas other vascular plants, such as ferns, reproduce by means of spores and so need moisture to develop, some seed plants can survive and reproduce in extremely arid conditions.

After the female part of the flower is pollinated, pollen grains attempt to travel into the ovary by creating a path called "pollen tube." The pollen tube does not directly reach the ovary in a straight line. It travels near the skin of the style and curls to the bottom of the ovary, then near the receptacle, it breaks through the ovule and reaches the ovum to fertilize it. After being fertilized, the ovary starts to swell and becomes a fruit.

With multi-seeded fruits, multiple grains of pollen are necessary for syngamy with each ovule. The process is easy to visualize if one looks at corn silk, which is the female flower of corn. Pollen from the tassel falls on the sticky external portion of the silk, then pollen tubes grow down the silk to the attached ovule. The dried silk remains inside the husk of the ear as the seeds mature, so one can carefully remove the husk to show the floral structures. The development of the flesh of the fruit is proportional to the percentage of fertilized ovules. In all, the book examines carefully, and with many vivid details. It tries to integrate the perspective of the social and technical scientis with the way the look at reality.

Editor

Contents

Chapter 1

Soil Fertility

As the years went by man became less of a wanderer and more of a settler. Families, clans, and villages developed, and with them came the development of the skill we call agriculture. It is generally agreed that one area in the world that shows evidence of a very early civilization is Mesopotamia, situated between the Tigris and Euphrates rivers in what is now Iraq. Writings dating back to 2500 B.C. mention the fertility of the land. It is recorded that the yield of barley was 86-fold and even 300-fold in some areas, which means, of course, that for every unit of seed planted 86 to 300 units were harvested.

Herodotus, the Greek historian, reporting on his travels through Mesopotamia some 2,000 years later, mentions the phenomenal yields obtained by the inhabitants of this land. The high production was probably the result of a well-developed irrigation system and soil of high fertility, attributable in part to annual flooding by the river. Theophrastus, writing around 300 B.C., referred to the richness of the Tigris alluvium and stated that the water was allowed to remain on the land as long as possible so that a large amount of silt might be deposited. In time man learned that certain soils would fail to produce satisfactory yields when cropped continuously. The practice of adding animal and vegetable manures to the soil to restore fertility probably developed from such observations, but how or when fertilisation actually began is not known.

Greek mythology, however, offers one picturesque explanation: Augeas, a legendary king of Elis, was famous for his stable, which contained 3,000 oxen. This stable had not been cleaned for thirty years. King Augeas contracted with Hercules to clean the stable out and agreed to give him 10 per cent of the cattle in return. Hercules is said to have accomplished this task by turning the River Alpheus through the stable, thus carrying away the accumulated filth and presumably depositing it on the adjacent land. Augeas then refused payment for this service, whereupon a war ensued, and Augeas was put to death by Hercules.

In the Greek epic poem The Odyssey, attributed to the blind poet Homer, who is thought to have lived between 900 and 700 B.C., the manuring of vineyards by the father of Odysseus is mentioned. The manure heap, which would suggest its systematic collection and storage, is also referred to. Argos, the faithful hound of Odysseus, was described as lying on such a heap when his master returned after an absence of twenty years. Having recognized his master, Argos wagged his tail feebly and "went down into the blackness of death." These writings suggest that manuring was an agricultural practice in Greece nine centuries before the birth of Christ. Xenophon, who lived between 434 and 355 B.C., observed that "the estate has gone to ruin" because "someone didn't know it was well to manure the land." And again, ". . . there is nothing so good as manure."

Theophrastus recommended the abundant manuring of thin soils but suggested that rich soils be manured sparingly. He also endorsed a practice considered good today-the use of bedding in the stall. He pointed out that this would conserve the urine and bulk and that the humus value of the manure would be increased. It is interesting to note that Theophrastus suggested that plants with high nutrient requirements also had a high water requirement. The truck gardens and olive groves around Athens were enriched by sewage from the city. A canal system was used, and there is evidence of a device for regulating the flow.

It is believed that the sewage was sold to farmers. The ancients also fertilised their vineyards and groves with water that contained dissolved manure. Manures were classified according to their richness or concentration. Theophrastus. for example, listed them in the following order of decreasing value: human, swine, goat, sheep, cow, oxen, and horse.

Later Varro, an early writer on Roman agriculture, developed a similar list but rated bird and fowl manure as superior to human excrement. Columella recommended the feeding of snail clover to cattle because he felt that it enriched the excrement. Not only did the ancients recognize the merits of manure, but they also observed the effect that dead bodies had on increasing the growth of crops. Archilochus made such an observation around 700 B.C., and the Old Testament records are even earlier. In Deuteronomy it is directed that the blood of animals should be poured on the ground. The increase in the fertility of land that has received the bodies of the dead has been acknowledged down through the years.

The value of green-manure crops, particularly legumes, was also soon recognized. Theophrastus noted that a bean crop was plowed under by the farmers of Thessaly and Macedonia. He observed that even when thickly sown and large amounts of seed were produced the crop enriched the soil. Xenophon. around 400 B.C., recommended spring plowing because "the land is more friable then" and "the grass turned up is long enough at that season to serve as manure, but not having shed seed, it will not grow." He also pointed out that "every kind of vegetation, every kind of soil in stagnant water turns into manure." Cato (234-149 B.C.) suggested that poor vineyard land be interplanted with a crop of acinum. It is not known what this crop is, but it was not allowed to go to seed, and the implication is that it was turned under. He also said that the best legumes for enriching the soil were field beans, lupines, and vetch.

Lupine was quite popular with the ancients. Columella listed numerous legume crops, including lupines, vetch, lentils, chick peas, clover, and alfalfa, that were satisfactory for soil improvement. Many of the early writers agreed, however, that lupine was the best general-purpose green-manure crop because it grew well under a wide range of soil conditions, furnished food for man and beast, was easy to seed, and quick to grow. Virgil (70-19 B.C.) advocated the application of legumes, as indicated in the following passage:

> Or, changing the season, you will sow there yellow wheat, whence before you have taken up the joyful pulse, with rustling pods, or the vetch's slender offspring and the bitter lupine's brittle stalks, and rustling grove.

The use of what might now be called mineral fertilisers or soil amendments was not entirely unknown to the ancients. Theophrastus suggested the mixing of different soils as a means of "remedying defects and adding heart to the soil." This practice may have been beneficial from several standpoints. The addition of fertile soil to infertile soil could lead to increased fertility, and the practice of mixing one soil with another may have provided better inoculation of legume seed on some fields. Again, the mixing of coarse-textured soils with those of fine texture or vice versa may have caused an improvement in the water and air relations in the soils of the fields so treated.

The early dwellers of Aegina dug up marl and applied it to their land. The Romans, who learned this practice from the Greeks and Gauls, even classified the various liming materials and recommended that one type be applied to grain and another to meadow. Pliny (62-113) stated that lime should be spread thinly on the ground and that one treatment was "sufficient for many years, though not 50." Columella also recommended the spreading of marl on a gravelly soil and the mixing of gravel with a dense calcareous soil. The Bible records the value of wood ashes in its reference to the burning of briars and bushes by the Jews, and Xenophon and Virgil both report the burning of stubble to clear fields and destroy weeds. Cato advised the vine keeper to burn prunings on the spot and to plow in the ashes to enrich the soil.

Pliny states that the use of lime from lime kilns was excellent for olive groves, and some farmers burned manure and applied the ashes to their fields. Columella also suggested the spreading of ashes or lime on lowland soils to destroy acidity. Saltpeter, or potassium nitrate, was mentioned both by Theophrastus and Pliny as useful for fertilising plants and is referred to in the Bible in the book of Luke. Brine was mentioned by Theophrastus. Apparently recognizing that palm trees required large quantities of salt, early farmers poured brine around the roots of their trees.

Even as soil scientists today are searching for methods of predicting the fitness of soil for the production of crops, so did the minds of the early agricultural philosophers and writers turn to such matters. Virgil, for example, believed that soil that was "blackish and fat under the deep pressed share, and whose mold is loose and crumbling is generally best for corn." Virgil wrote on the soil characteristic known today as bulk density. His advice on determining this property was

> . . . first, you shall mark out a place with your eye, and order a pit to be sunk deep in solid ground, and again return all the mold into its place, and level with your feet the sands at top. If they prove deficient, the soil is loose and more fit for cattle and bounteous vines: but if they deny the possibility of returning to their places, and there be an over plus of mold after the pit is filled up, it is a dense soil; expect reluctant clods and stiff ridges, and give the first plowing to the land with sturdy bullocks.

Pliny apparently held a different view of the merits of the test proposed by Virgil. He stated that soil thrown back into the hole from which it was dug will never fill the hole, and that by this method it is impossible to form any opinion concerning its density or thinness. Later, Columella, writing on ways of testing the soil to determine its physical suitability for agricultural use, set down an approach different from that of Virgil. He suggested that a trench be dug and that the soil removed therefrom be returned to it. If the hole were then not filled completely, the soil was lean, but if some soil were left out the soil was fat. Virgil describes another means which might today be considered the prototype of a chemical soil test.

But saltish ground, and that which is accounted bitter, where corn can never thrive, will give proof of this effect. Snatch from the smoky roofs bushels of close woven twigs and the strainers of the wine press. Hither let some of that vicious mold, and sweet water from the spring, be pressed brimful; be sure that all the water will strain out and bid drops pass through the twigs. But the taste will clearly make discovery; and in its bitterness will distort the wry faces of the tasters with the sensation. Columella also suggested a taste test to measure the degree of acidity and salinity of soils, and Pliny stated that the bitterness of soils might be detected by the presence of black and underground herbs.

Pliny wrote that "among the proofs of the goodness of soil is the comparative thickness of the stem in corn," and Columella stated simply that the best test for the suitability of land for a given crop was whether it would grow. Many of the early writers believed that the color of the soil was a criterion of its fertility. The general idea was that black soils were fertile and light or gray soils infertile. Columella disagreed with this viewpoint, pointing to the infertility of the black marshland soils and the high fertility of the light-colored soils of Libya. He felt that such factors as structure, texture, and acidity were far better guides to an estimation of soil fertility. Much of the early writing regarding soil fertility consisted largely of descriptions of farm practices. There seems to be little evidence of an experimental approach to farm problems, but many of these manuscripts do reflect a rather keen comprehension of certain of the factors now known to affect plant growth.

Some of the practices described in these early writings are as sound today as they were then. Many more are not, but it was centuries before man began to work intelligently toward a solution of the agricultural problems with which he was confronted. The age of the Greeks from perhaps 800 to 200 B.C. was indeed a Golden Age. Many of the men of this period reflected genius that was unequaled, or at least not permitted expression, for centuries to come. Their writings, their culture, their agriculture were copied by the Romans, and the philosophy of many of the Greeks of this period dominated the thinking of men for more than 2,000 years.

SOIL FERTILITY AND THE FIRST EIGHTEEN CENTURIES A.D.

After the decline of Rome there were few contributions to the development of agriculture until the publication of Opus ruralium commodorium, a collection of local agricultural practices, by Pietro de Crescenzi (1230-1307). De Crescenzi is referred to by some as the founder of modern agronomy, but his manuscript seems to be confined to the work of writers from the time of Homer. His contribution was largely that of summarizing the material. He did, however, suggest an increase in the rate of manuring over that in use at the time.

After the appearance of de Crescenzi's work little was added to agricultural knowledge for many years, although Palissy in 1563 is credited with the observation that the ash content of plants represented the material they had removed from the soil. Around the beginning of the seventeenth century Francis Bacon (1561-1624) suggested that the principal nourishment of plants was water. He believed that the main purpose of the soil was to keep the plants erect and to protect them from heat and cold and that each plant drew from the soil a substance unique for its own particular nourishment. Bacon maintained further that continued production of the same type of plant on a soil would impoverish it for that particular species.

During this same period Jan Baptiste van Helmont (1577-1644), a Flemish physician and chemist, reported the results of an experiment which he believed proved that water was the sole nutrient of plants. He placed 200 lb. of soil into an earthen container, moistened the soil, and planted a willow shoot weighing 5 lb. He carefully shielded the soil in the crock from dust, and only rain or distilled water was added. After a period of five years, van Helmont terminated the experiment. The tree weighed 169 lb. and about 3 oz. He could account for all but about 2 oz. of the 200 lb. of soil originally used. Because he had added only water, his conclusion was that water was the sole nutrient of the plant, for he attributed the loss of the 2 oz. of soil to experimental error.

Of course, it is now known that both carbon dioxide (CO_2) and minerals from the soil are required for the nutrition of plants. It should be remembered, however, that this work was done at a time before anything was known about mineral nutrition or photosynthesis. Van Hel-mont's work and his erroneous conclusions were actually valuable contributions to our knowledge, for, even though they were wrong, his conclusions stimulated later investigations, the results of which have led to a much better understanding of plant nutrition.

The work of van Helmont was repeated several years later by no less a figure than Robert Boyle of England. Boyle is probably best known for expressing the relation of the volume of a gas to its pressure. He was also interested in biology and a great believer in the experimental approach to the solution of problems dealing with science. He believed that observation was the only road to truth. Boyle confirmed the findings of van Helmont, but he went one step farther. As a result of the chemical analyses he performed on plant samples, he stated that plants contained salts, spirits, earth, and oil, all of which were formed from water. About this same time J. R. Glauber, a German chemist, suggested that saltpeter (KNO_3) and not water was the "principle of vegetation." He collected the salt from soil under the pens of cattle and argued that it must have come from the droppings of these animals.

He further stated that because the animals ate forage the saltpeter must have come originally from the plants. When he applied this salt to plants and observed the large increases in growth it produced, he was convinced that soil fertility and the value of manure were due entirely to saltpeter. John Mayow, an English chemist, supported the viewpoint of Glauber. Mayow estimated the quantities of niter in the soil at various times during the year and found it in its greatest concentration in the spring. Failing to find any during the summer, he concluded that the saltpeter had been absorbed, or sucked up, as he put it, by the plant during its period of rapid growth. Experimental techniques were quite crude at this time.

Even with the contributions made by Mayow, Glauber, Boyle, and Bacon, there was really little work that could, by present-day standards, be termed good research. About 1700, however, a study was made which was outstanding and which represented a considerable advance in the progress of agricultural science.

An Englishman by the name of John Woodward, who was acquainted with the work of Boyle and van Helmont, grew spearmint in samples of water he had obtained from various sources: rain water, river water, sewage water, and sewage water plus garden mold. He carefully measured the quantity of water transpired by the plants and recorded the weight of the plants at the beginning and end of the experiment. He found that the growth of the spearmint was proportional to the amount of impurities in the water and concluded that terrestrial matter, or earth, rather than water, was the principle of vegetation.

Although his conclusion is not correct in its entirety, it represents an ad-vance in knowledge, and his experimental technique was considerably better than any that had been used before. There was much understandable ignorance concerning the nutrition of plants during this period. Many quaint ideas came into being and "bode their hour or two and went their way." Not the least of these ideas were introduced by another enterprising Englishman, Jethro Tull (1674-1741). Tull was educated at Oxford, which is generally considered a bit unusual for a person of agricultural leanings. He appears to have been interested in politics, but ill health forced his retirement to the farm. There he carried out numerous experiments, most of which dealt with cultural practices. He believed that the soil should be finely pulverized to provide the "proper pabulum" for the growing plant.

According to Tull, the soil particles were actually ingested through openings in the plant roots. The pressure caused by the swelling of the growing roots was thought to force this finely divided soil into "the lacteal mouths of the roots," after which it entered the "circulatory system" of the plant. Tull's ideas about plant nutrition were, to say the least, a bit odd. However, his experiments led to the development of two valuable pieces of farm equipment, the drill and the horse-drawn cultivator. His published book, Horse Hoeing Husbandry, was long considered an authoritative text in English agricultural circles. Around 1762 John Wynn Baker, a Tull adherent, established an experimental farm in England, the purpose of which was the public exhibition of the results of experiments in agriculture.

Baker's work was praised later by Arthur Young who, however, admonished his readers to beware of giving too much credit to calculations based on the results of only a few years' work, an admonition that is as timely today as it was when originally made.

One of the more famous of the eighteenth-century English agriculturists was Arthur Young. Young conducted pot tests to find those substances that would improve the yield of crops. He grew barley in sand to which he added such materials as charcoal, train oil, poultry dung, spirits of wine, niter, gunpowder, pitch, oyster shells, and numerous other materials. Some of the materials produced plant growth, others did not. Young, a prolific writer, published a work entitled Annals of

Agriculture, in forty-six volumes, which was highly regarded and made a considerable impact on English agriculture.

Many of the agricultural writings of the seventeenth and eighteenth centuries reflected the idea that plants were composed of one substance, and most of the workers during this period were searching for this principle of vegetation. Around 1775, however, Francis Home stated that there was not only one principle but probably many, among which he included air, water, earth, salts, oil, and fire in a fixed state. Home felt that the problems of agriculture were essentially those of the nutrition of plants. He carried out pot experiments to measure the effects of different substances on plant growth and made chemical analyses of plant materials. His work was considered to be a valuable stepping stone in the progress of scientific agriculture.

The idea that plants contained fire in a fixed state, or phlogiston, as it was called, lingered in the minds of men for many years. There was also the belief that organic material or humus was taken in directly by plants and that it constituted their principal nutrient. This idea persisted down through the years. It was difficult to dispel because the results of chemical analyses had shown that plants and humus contained essentially the same elements. Too, the process of photosynthesis had not yet been discovered. Joseph Priestley's observation around 1775 that sprigs of mint "purified" air led him to suggest that plants reversed the effect of breathing.

At the time of this observation oxygen had not been discovered. Later, when he did discover this gas, he failed to recognize its relation to plants. The discovery of oxygen by Priestley was the keystone to a number of other discoveries that went far toward unlocking the mystery of plant life. Jan Ingenhousz showed that the purification of air took place in the presence of light, but in the dark the air was not purified. Coupled with this discovery was the statement by Jean Senebier , a Swiss natural philosopher and historian, that the increase in the weight of van Helmont's willow tree was the result of air!

PROGRESS DURING THE NINETEENTH CENTURY

These discoveries stimulated the thinking of Theodore de Saussure, whose father was acquainted with the work of Senebier. He attacked two of the problems on which Senebier had worked-the effect of air on plants and the origin of salts in plants. As a result, de Saussure was able to demonstrate that plants absorbed oxygen and liberated carbon dioxide, the central theme of respiration.

In addition, he found that plants would absorb carbon dioxide with the release of oxygen in the presence of light. If, however, plants were kept in an environment free of carbon dioxide, they died. DeSaussure concluded that the soil furnishes only a small fraction of the nutrients needed by plants, but he demonstrated that it does supply both ash and nitrogen. He effectively dispelled the idea that plants spontaneously generate potash and stated further that the plant root does not behave as a mere filter. Rather the membranes are selectively permeable, allowing for a more rapid entrance of water than of salts. He also showed the differential absorption of salts and the inconstancy of plant composition, which varies with the nature of the soil and the age of the plant.

De Saussure's conclusion that the carbon contained by plants was derived from the air did not meet with immediate acceptance by his colleagues. No less a figure than Sir Humphrey Davy, who published his work The Elements of A gricultural Chemistry about 1813, stated that although some plants may have received their carbon from the air the major portion was taken in through the roots. Davy was so enthusiastic in this belief that he recommended the use of oil as a fertiliser because of its carbon and hydrogen content. The middle of the nineteenth to the beginning of the twentieth century was a time during which much progress was made in the understanding of plant nutrition and crop fertilisation.

Among the men of this period whose contributions loom large was Jean Baptiste Boussingault, a widely traveled French chemist who established a farm in Alsace on which he carried out field-plot experiments. Boussingault employed the careful techniques of de Saussure in weighing and analyzing the manures he added to his plots and the crops he harvested. He maintained a balance sheet which showed how much of the various plant-nutrient elements came from rain, soil, and air, analyzed the composition of his crops during various stages of growth, and determined that the best rotation was that which produced the largest amount of organic matter in addition to that added in the manure. Boussingault is called by some the father of the field-plot method of experimentation. Although several of the agricultural scientists of this period recognized the value of de Saussure's observations, the old humus theory still had many adherents. It was such a natural theory that it was difficult to dispel.

Many must have felt then, as some do today. that the decay of plant and animal material gives rise to products that are best suited for the nutrition of growing plants. Justus von Liebig, a German chemist, very effectively deposed the humus myth. The presentation of his paper at a prominent scientific meeting jarred the conservative thinkers of the day to such an extent that only a few scientists since that time have dared to suggest that the carbon contained in plants comes from any source other than carbon dioxide. Liebig made the following statements:

- Most of the carbon in plants comes from the carbon dioxide of the atmosphere.
- Hydrogen and oxygen come from water.
- The alkaline metals are needed for the neutralization of acids formed by plants as a result of their metabolic activities.
- Phosphates are necessary for seed formation.
- Plants absorb everything indiscriminately from the soil but excrete from their roots those materials that are nonessential.

Not all of Liebig's ideas, of course, were correct. He thought that acetic acid was excreted by the roots. He also believed that the NH_4^+ form of nitrogen was the one absorbed and that plants might obtain this compound from soil, manure, or air.

Liebig firmly believed that by analyzing the plant and studying the elements it contained one could formulate a set of fertiliser recommendations based on these analyses. It was also his opinion that the growth of plants was proportional to the amount of mineral substances available in the fertiliser. He eventually developed the law of the minimum, which, in effect, says that the growth of plants is limited by the plant-nutrient element present in the smallest quantity, all others being present in adequate amounts.

This concept dominated the thinking of agricultural workers for a long time thereafter. Liebig manufactured a fertiliser based on his ideas of plant nutrition. The formulation of the mixture was perfectly sound, but he made the mistake of fusing the phosphate and potash salts with lime. As a result the fertiliser was a complete failure. Nonetheless. the contributions that Liebig made to the advancement of agriculture were monumental, and he is perhaps quite rightly recognized as the father of agricultural chemistry. Following on the heels of Liebig's now famous paper was the establishment in 1843 of an agricultural experiment station at Rothamsted, England. The founders of this institution were J. B. Lawes and J. H. Gilbert. Work here was conducted along the same lines as that carried out earlier by Boussingault in France.

Lawes and Gilbert did not believe that all of the maxims set down by Liebig were correct. Twelve years after the founding of the station they settled the following points:

— Crops require both phosphorus and potash, but the composition of the plant ash is no measurement of the amounts of these constituents required by the plant.
— Nonlegume crops require a supply of nitrogen. Without this element, no growth will be obtained, regardless of the

quantities of phosphorus and potassium present. The amount of ammonia nitrogen contributed by the atmosphere is insufficient for the needs of crops.

— Soil fertility could be maintained for some years by means of chemical fertilisers.

— The beneficial effect of fallow lies in the increase in the availability of nitrogen compounds in the soil.

For a long time, and even today in some sections, farmers were reluctant to believe that fertility could be maintained by the use of chemical fertilisers alone. The early work at Rothamsted proved conclusively, however, that it can be done. When soils become depleted in spite of the use of commercial fertilisers, it is largely because of inadequate amounts of these materials and improper soil management rather than the mineral fertiliser per se.

Work at the Rothamsted station has continued on through the years, and today this station is recognized as one of the world's leading centers of agricultural research. A discovery of considerable importance was made in England in 1852. Thomas Way, observing that a Yorkshire farmer was able to reduce ammonia losses from manure by the addition of soil, first demonstrated the phenomenon known as cation exchange. However, its tremendous significance was not immediately recognized, even by the great Liebig. The value of some of the early results from the Rothamsted experiments was recognized by Georges Ville, a Frenchman of Vincennes. Ville maintained that the use of chemical fertilisers was the only method of supporting soil fertility. He made recommendations for the fertilisation of crops based on the results of field trials and drew up a simple scheme of plot tests that could be used by farmers to determine for themselves just what fertilisers were needed for their crops.

The problem of soil and plant nitrogen remained unsolved. Several workers had observed the unusual behavior of legumes.

In some instances they grew well in the absence of added nitrogen, whereas in others no growth was obtained. Nonlegumes, on the other hand, always failed to grow when there was insufficient nitrogen in the soil. In 1878 some light was thrown on the confusion by two French bacteriologists, Theodore Schloessing and Alfred Muntz. These scientists purified sewage water by passing it through a filter made of sand and limestone. They analyzed the filtrate periodically, and for twenty-eight days only ammonia nitrogen was detected. At the end of this time nitrates began to appear in the filtrate.

Schloessing and Muntz found that the production of nitrates could be stopped by adding chloroform and that it could be started again by adding a little fresh sewage water. They concluded that nitrification was the result of bacterial action. The results of these experiments were applied to the soils by Robert Warrington of England. He showed that nitrification could be stopped by carbon disulfide and chloroform and that it could be started again by adding a small amount of unsterilized soil. He also demonstrated that the reaction was a two-step phenomenon, the ammonia first being converted to nitrites and the nitrites subsequently to nitrates.

Warrington was unable to isolate the organisms responsible for nitrification. This task remained for S. Winogradsky, who effected the isolation by the use of a silica-gel plate, rather than the usual agar medium, because the organisms are autotrophic and obtain their carbon from the carbon dioxide of the atmosphere. As to the erratic behavior of legume plants with respect to nitrogen, two Germans, Hellriegel and Wilfarth, in 1886 concluded that bacteria must be present in the nodules attached to legume roots. Further, these organisms were believed to assimilate gaseous nitrogen from the atmosphere and to convert it to a form that could be used by higher plants. This was the first specific information regarding nitrogen fixation by legumes.

Hellriegel and Wilfarth based their arguments on observations made in certain of their experiments. They did not, however, isolate the responsible organisms. This was later done by M. W. Beijerinck, who called the organism Bacillus radicicola.

Although most of the advances made in agriculture during the eighteenth century were accomplished on the Continent, a few early American contributions were sufficiently significant to mention. In 1733 James E. Oglethorpe established an experimental garden on the bluffs of the Savannah River, which is the present site of the city of Savannah, Georgia. This garden was devoted to the production of exotic food crops and is said to have been a place of beauty while it was maintained. Interest was lost in it, however, and it soon ceased to exist, but because it was largely the result of British interests it probably cannot be truly considered an "American" undertaking.

Benjamin Franklin demonstrated the value of gypsum. On a prominent hillside he applied gypsum to the land in a pattern which outlined the words, "This land has been plastered." The increased growth of pasture in the area to which the gypsum had been applied served as an effective demonstration of its fertiliser value. In 1785 a society was formed in South Carolina which had among its objectives the setting up of an experimental farm. Eleven years later President Washington, in his annual message to the Congress, pleaded for the establishment of a national board of agriculture.

Some of the most important contributions to early American agriculture were made by Edmond Ruffin of Virginia from about 1825 to 1845. He is believed to have been one of the first to use lime on humid-region soils to replace fertility elements lost by crop removal and leaching. Ruffin was a careful observer, a studious reader, and possessed of a keen and inquiring mind. Although his use of lime to bolster the yields of crops was known to the ancients, it was apparently a new experience in a new America. It was not until 1862 that the Department of

Agriculture was established, and, in the same year, the Morrill Act provided for state colleges of agriculture and mechanical arts. The first organized agricultural experiment station, set up in 1875 at Middletown, Connecticut, was supported by state funds.

In 1877 North Carolina established a sim-ilar station, followed closely by New Jersey, New York, Ohio, and Massachusetts. In 1888 the Hatch Act called for the setting up of state experiment stations to be operated in conjunction with the Land-Grant colleges, and an annual grant of $15,000 was made available to each state for their support. Although much of the early experimental work was largely demonstration, a "scientific" approach to agricultural problems was gradually developed in this country. The idea that plants excreted weak acids was firmly fixed in the minds of many, and a method for determining the fertility status of soils by weak acid extraction came into being.

Although such treatment produced results that were of little absolute value, it was found, when large numbers of soils were studied, that the amounts of nutrients extracted generally were correlated with crop yields. The suggestion was advanced that mineral elements were present in the soil in two rather well-defined forms: in the undecomposed primary minerals and in secondary minerals known as zeolites. These elements were thought to be held in available form by the zeolites. The idea of extracting soils with acids to determine their fertility status persisted, and E. W. Hilgard found that the maximum solubility of soil minerals in hydrochloric acid (HCl) was obtained when the acid was at a specific gravity of 1.115, which happens to correspond to the concentration of the acid obtained on prolonged boiling. Hilgard attached particular significance to this fact. The strong acid digestion became quite popular, and numerous analyses of soils were made by this method. It was later shown that there was little foundation for assuming that this technique would give data of predictive value, and its use was discontinued.

Two early workers who contributed much to the development of interest in soil fertility in the United States were Milton Whitney and C. G. Hopkins. Around the beginning of the twentieth century these two men engaged in a controversy which attracted nationwide attention and which, in fact, became quite bitter. Whitney maintained that the total supply of nutrients in soils was inexhaustible and that the important factor from the standpoint of plant nutrition was the rate at which these nutrients entered into the soil solution. Hopkins, on the other hand, felt that this philosophy would lead to soil depletion and a serious decline in crop production. He made a survey of the soils of Illinois and reduced soil fertility to a system of bookkeeping.

As a result of these exhaustive studies, he concluded that Illinois soils required the addition of only lime and phosphate. So effectively did he preach this doctrine that the use of lime and rock phosphate in a corn, oats, and clover rotation was a continuous practice in this state for many years. The controversy between Whitney and Hopkins finally waned. Whitney's ideas were shown to be at least partly incorrect, but the argument, though bitter, did much to stimulate the thinking of agricultural scientists of the period. Just after the turn of the twentieth century most experiment stations had established field plots which showed the remarkable benefits of fertilisation.

As a result of these experiments, the major problem of soil fertility could be delimited in a broad way. It was shown, for example, that there was a widespread need for phosphatic fertilisers, that potassium was generally lacking in the coastal plains regions, and that nitrogen was particularly deficient in the soils of the South. The soils east of the Mississippi were generally acid and needed lime, whereas those west of the river were as a rule fairly well supplied with calcium. Even though the broad outline of the fertility status of soils in the United States had been fairly well defined, it was soon apparent that blanket fertiliser recommendations based on such knowledge could not be made.

Each farm required individual attention, as did each field within that farm. The interest in soil tests sprang up anew. During the last thirty years much headway has been made toward an understanding of the problems of soil fertility. To enumerate the men whose contributions have advanced our knowledge would require far more space than is available here. These advances have not been the work of scientists of any one country.

The English, who began around 1600, have continued to make great strides. The agricultural workers of France, Germany, Scandinavia, Russia, Canada, Australia, New Zealand, as well as the United States and other countries, have unraveled many problems which have held back the progress of the science. The fruits of these studies are everywhere apparent. for agricultural production in advanced countries is higher today than it has ever been, and the free world as a whole is generally better fed, clothed, and housed than at any time in the past. This could not be so if the production of crops today were at the level of Europe in the Dark Ages, when the average yield of grain was 6 to 10 bu./A.

Progress in agriculture depends on research of a high caliber. For every problem solved by the scientist today, many more are raised. Agricultural scientists must delve into questions of a fundamental nature, questions that deal more with the why of things than with the what. Certainly the advances made toward the end of the nineteenth century and in the twentieth have been largely responsible for the present state of our learning.

Chapter 2

Plant Nutrition and Fertilisers

Plant nutrition refers to the needs and uses of the basic chemical elements in the plant. Fertilisation is the term used when these materials are supplied to the environment around the plant. Chemical change may occur before a plant nutrient supplied in a fertiliser can be taken up and used by the plant. Plants need 16 elements for normal growth. Carbon, hydrogen and oxygen are found in air and water.

Nitrogen, potassium, magnesium, calcium, phosphorus and sulfur are found in soil. These six elements are used in relatively large amounts by plants and are called macronutrients. There are seven other elements known as micronutrients or trace elements that are used in much smaller amounts. These essential micronutrients are found in soil and include iron, zinc, molybdenum, boron, copper, manganese and chlorine. All 16 elements, both macronutrients and micronutrients, are essential for plant growth. Other elements also associated with plant growth include cobalt, selenium and silicon.

Nutrient Absorption

Most of the nutrients that a plant needs are dissolved in water and then absorbed by the roots. Ninety-eight percent of these plant nutrients are absorbed from the soil solution, and only about 2 percent are actually extracted from soil particles by the roots. Most of the nutrient elements are absorbed as charged ions or pieces of molecules (which are the smallest particle of a substance that can exist and still retain the characteristics of the substance). Ions may be positively charged cations or negatively charged anions. Positive and negative are equally paired so that there is no overall charge. For example, nitrogen may be absorbed as nitrate (NO_3^-), which is an anion with one negative charge. The potassium ion (K^+) is a cation with one positive charge. Potassium nitrate (KNO_3^-) has one nitrate ion and one potassium ion.

However, calcium nitrate ($Ca(NO_3^-)_2$) has two nitrate ions and one calcium ion because the calcium cation had two positive charges. The balance of ions in the soil is very important. Just as ions of cpposite charges attract each other, ions of similar charges compete for chemical interactions and reactions in the environment. Some ions are more active than others and compete better. For example, both calcium (Ca^{++}) and magnesium (Mg^{++}) are cations with two charges. Magnesium with a smaller radius is more active and can compete better. If both are in competition to be absorbed, the magnesium will be absorbed.

Although a soil test may indicate that there is sufficient calcium in the soil, the plant may still exhibit a calcium deficiency because of an excess of the more active magnesium. Therefore, what may be expressed as a deficiency in one micronutrient may actually be caused by an excess of another. For ions to be easily absorbed, they must first be dissolved in the soil solution. Some combinations of ions are easily dissolved, such as potassium nitrate. When other ions combine, they may precipitate or fall out of solution and thus become unavailable to the plant. Many of the micronutrients form complex combinations with phosphorus and calcium and precipitate out of the soil solution so the nutrients cannot be easily taken up by the plant.

The pH, which is a measurement of acidity or alkalinity, greatly affects these chemical reactions. If soil pH is extremely high (alkaline), many of the micronutrients precipitate out of solution and are unavailable to the plant. When soil pH is extremely low (acidic), some of the micronutrients become extremely soluble and ion levels may become high enough to injure the plant. The effect of pH varies with the ion, the types of ions in the soil and the type of soil. Therefore, not only is the amount of the nutrient important, but also soil pH.

Water and oxygen in soil must also be available. Since nutrients are taken up with water, there must be sufficient water for the plant to absorb. Water is taken into the plant both passively and actively. Water that is taken in passively requires no energy output by the plant. It flows through the plant due to differences in concentration between the soil solution and the liquid within the cell. Water that is actively absorbed requires energy from the plant. If there is no oxygen available, sugar cannot be burned (oxidized) to produce energy and, therefore, water and nutrients cannot be absorbed.

Anything that lowers or prevents the production of sugars in the leaves can lower nutrient absorption. If the plant is under stress due to low light or extremes in temperature, nutrient deficiency problems may develop. The growth stage or how actively the plant is growing may also affect the amount of nutrients absorbed. Many plants go into a rest or dormancy during part of the year. During this rest period, few nutrients are absorbed. Plants may also absorb different nutrients just as flower buds begin to develop.

Nutrients transported from the root to the cell by the vascular system move into the cell across a cell membrane. There are three different ways this happens. First, an entire molecule or ion pair may move through the membrane. If the cell is using energy or active transport to absorb the ions, then only one of the ions of the pair is pulled into the cell. The other will follow to keep the number of positive and negative charges even. Most anions (negative ions) are actively absorbed. The second way of keeping the charges inside the cell balanced and absorbing a new ion is to exchange one charged ion for another of the same charge.

The hydrogen ion (H^+) is often released from the cell so that the cell can absorb another positive ion such as the potassium ion (K^+). Since this is a simple exchange or passive, absorption energy may not be required. Cations or positive ions may be passively absorbed by this method. Both of the methods mentioned above may be passive or active. However, the third method, the carrier system, is always active absorption and requires energy. Scientists have discovered that within the cell membrane there are specialized chemicals that act as carriers.

The carrier, through chemical changes, attracts an ion outside the cell membrane and releases it inside the cell. Once the ion is inside the cell, it is attached to other ions so that it does not move out of the cell. Complex chemical reactions are involved in the entire process. Although nutrients can be passively absorbed, research has shown that active absorption must take place if the plant is to grow and be healthy.

FOLIAR ABSORPTION

Under normal growing conditions, plants absorb most nutrients from the soil, with the exceptions of carbon, hydrogen and oxygen. Some nutrients can also be absorbed by the leaves if a dilute solution is sprayed on the leaves. Factors affecting absorption by the cell are still important because the nutrient must enter the cell to be used by the plant. Care must be taken that the nutrient concentration is not too high or the leaf will be injured. Also, the leaf is covered by a thin layer of wax called the cuticle. The nutrient must pass around or through the cuticle before it can enter the cell.

FERTILISERS

Fertilisers are materials containing plant nutrients that are added to the environment around the plant. Generally, they are added to water or soil, but some can also be added to the air or sprayed on leaves. Fertilisers are not plant food! Although it is common for many fertilisers to be called plant food, it is a misnomer. Plants produce their own food using water, carbon dioxide and energy from the sun. This food (sugars and carbohydrates) is combined with the plant nutrients to produce protein, enzymes, vitamins and other elements essential to plant growth.

DIFFERENT TYPES OF FERTILISERS

Of the 16 essential elements for plant growth, nitrogen, phosphorus and potassium are considered fertiliser macronutrients because plants require them in larger quantities for maximum growth.

Common incomplete fertilisers or farm type fertilisers.

	% Nitrogen	% Phosphorus	% Potassium
Ammonium nitrate	33	0	0
Ammonium sulfate	21	0	0
Mono-ammonium phosph	11	52	0
Muriate of potash (potassium chloride)	0	0	60
Potassium sulfate	0	0	52
Super phosphate	0	20	0
Triple super phosphate	0	46	0
Urea	42-45	0	0
Diamonium phosphate	18	46	0
Slow Release			
Urea-formaldehyde	38	0	0
Sulfur-coated urea	22-17	0	0
IIBDU	31	0	0

Calcium, magnesium and sulfur are secondary macronutrients, but they are usually either present in sufficient quantities, or they are added coincidentally with other materials. The other essential elements, called micronutrients, are just as important but are required in smaller amounts. If plants lack any of these elements, they exhibit signs of nutrient deficiency.

FERTILISER ANALYSIS

The fertiliser analysis on the bag refers to how much of an element there is in a formulation based on percentage of weight. All fertilisers are labeled with three numbers that give the percentage by weight of nitrogen (N), phosphate (P_2O_5) and potash (K_2O). Often, to simplify matters, these numbers are said to represent nitrogen, phosphorus, and potassium, or N-P-K. We should remember that it is not N-P-K but N-P_2O_5-K_2O. For example, if we have a 100-pound bag of fertiliser labeled 10-10-10, there are 10 pounds of N, 10 pounds of P_2O_5 and 10 pounds of K_2O. To convert the P_2O_5 to actual phosphorus, multiply it by 0.43; to convert K_2O to actual potassium, multiply it by 0.83.

In this example, the other 70 pounds is filler or carrier. Filler or carrier helps to evenly spread the fertiliser and avoids burning plants with too much fertiliser. A 100-pound bag of fertiliser labeled 0-20-10 would have 0 pounds of N, 20 pounds of P_2O_5, 10 pounds of K_2O and 70 pounds of filler or carrier. A number of states have adopted a model label law for classifying fertilisers. The law establishes minimum levels of nutrients allowable and provides specific labeling requirements. To date, model label legislation has not met with total acceptance, so there are still differences from state to state as to what constitutes a fertiliser and the type of information required on labels.

Even so, the information contained on fertiliser labels has been well-standardized, and the consumer is protected by state laws requiring manufacturers to guarantee the claimed nutrients. The law only requires that the manufacturer guarantee what is claimed on the label, so in some cases, a fertiliser will contain secondary nutrients or micronutrients not listed on the label because the manufacturer does not want to guarantee their exact amounts. The gardener-consumer can rest assured that nutrients listed on the label are actually contained in the fertiliser. The best fertiliser analysis to use depends on many factors, such as what nutrients are needed, what the soil structure and chemistry are and the method of applying the fertiliser.

Complete and Incomplete Fertilisers

A fertiliser is said to be complete when it contains nitrogen, phosphorus and potassium. The manufacturers of commercial fertilisers are required to state nutrient amounts on the container as a guaranteed analysis. Examples of commonly used fertilisers are 10-10-10, 15-5-10, 20-10-5. An incomplete fertiliser will be missing one of the major components. The blending of incomplete fertilisers is used to make complete fertilisers. As in Example 1, if 100 pounds of urea (45-0-0) were combined with 100 pounds of triple super phosphate (0-45-0), and 100 pounds of muriate of potash (0-0-60), a fertiliser grade of 15-15-20 would result.

The fertiliser ratio indicates the proportion of nitrogen, phosphate and potash contained in the fertiliser. The specific fertiliser ratio needed depends on the soil nutrient level. For example, a 1-1-1 ratio (i.e., 10-10-10, 15-15-15, 20-20-20) is widely used at the time of lawn establishment, but established lawns generally respond better to fertiliser ratios high in nitrogen. Three of the most common complete fertilisers used by homeowners are 10-10-10, 19-5-9 and 15-5-10.

Purpose of Fertilisers

When shopping for fertiliser, you will find that some are packaged for certain uses or types of plants, such as camellia food, rhododendron and azalea food or rose food. The camellia and rhododendron or azalea fertilisers belong to an old established group, the acid plant foods. Some of the compounds used in these fertilisers are chosen because they have an acid reaction; they are especially beneficial to acid-loving plants in soil which is naturally neutral or alkaline. The other fertilisers packaged for certain plants do not have as valid a research background. Compare, for example, the fertiliser ratios of three different brands of rose food on any garden center shelf.

Other special purpose fertilisers, which continue to be marketed extensively, claim to promote increased blooming of vegetable and ornamental plants. These fertilisers contain high amounts of phosphorous. Although phosphorous is a macronutrient and needed by the plant for proper blooming and root growth, the addition of high doses of phosphorous does not guarantee increased blooming. Most New Mexico soils have phosphorous in them, and occasional additions of a complete lawn or garden fertiliser is generally adequate.

A soil test should be performed before purchasing any expensive, special-purpose fertiliser. It is not possible to make a blanket statement that one fertiliser is best for every area of the state. It is true that different plants use different nutrients at different rates. What is unknown is the reserve of nutrients already in the soil. This changes with every soil type and every location.

Slow-Release Fertilisers

Since plants can take up fertilisers continuously, it is beneficial to provide them with a balance of nutrients throughout their growth. Perhaps the most efficient way to achieve this is to apply a slow-release fertiliser, which is designed to release nutrients at the same rate they are taken up by the plants. Slow-release fertilisers contain one or more essential elements. These elements are released or made available to the plant over an extended time.

Slow-release fertilisers can be categorized by the way in which the fertiliser is released. The three major types of nutrient release include the following:

— materials that dissolve slowly,

— materials from which the nitrogen is released by microorganisms and

— granular materials with membranes made of resin or sulfur that control the rate of nutrient release from the granules into the soil.

	Advantages	Disadvantages
Slow-Release Fertiliser	• Fewer applications	• Unit cost is high
	• Low burn potential	• Availability limited
	• Release rates vary depending	• Release rate governed by on fertiliser characteristics factors other than plant need
	• Comparatively slow release rate	
Conventional Fertiliser	• Fast acting	• Greater bum potential
	• Some are acid	• Solidifies in the forming bag when w
	• Low cost	• Nitrogen leaches readily
Manures or Sewage	• Low bum potential	• Salt may be a problem
Sludge	• Relatively slow	• Bulky; difficult to handle
	• Contains micro-nutrients	• Odor
	• Conditions soil	• Expensive per pound of actu nutrient
		• Weed seeds a problem
		• Heavy metals may be prese

		sewage sludge (frequent in sludge from large cities)

These nutrient release systems may be used in combination with one another. Sulfur-coated urea is a slow-release fertiliser with a covering of sulfur around each urea particle. Different thicknesses of sulfur control the release rate of nitrogen, which becomes more rapid as temperature increases. Watering does not affect its release rate. When applied to the soil surface, sulfur-coated urea releases more slowly than if it is incorporated into the soil. This material generally costs less than other slow-release fertilisers and it supplies the essential element sulfur as well as Nitrogen.

When fertiliser products coated with multiple layers of resin come into contact with water, the layers swell and increase the pore size in the resin so that the dissolved fertiliser can move into the soil. Release rate depends on coating thickness, temperature and water content of the soil. There is often a large release of fertiliser during the first 2 or 3 days after application. Depending on the coating, release timing can extend up to 12 months.

Slow-release fertilisers need not be applied as frequently as other fertilisers, and higher amounts can be applied without danger of burning. Plants may use the nitrogen in slow-release fertilisers more efficiently than in other forms since it is being continually released over a longer period of time than in conventional fertilisers. Slow-release fertilisers are generally more expensive than other types. The real benefit, however, is the frequency of application, which is much lower than with conventional fertilisers. Urea formaldehyde and sulfur-coated urea have been used as turf fertiliser, while resin-coated fertilisers are predominantly used in container growing.

Caution should be used in applying slow-release fertilisers around trees or shrubs in the summer as they may keep the plant in growth late in summer. Late sea-son growth may not harden off completely and excessive winter damage may occur. Some mention should be made of the initials or designations W.I.N. and W.S.N. on fertiliser labels. These stand for water insoluble nitrogen and water soluble nitrogen, respectively. Water soluble nitrogen dissolves readily and is usually in a very simple form such as ammoniacal nitrogen (ammonia) or nitrate nitrogen.

Nitrogen that will not dissolve readily may exist in other forms in the fertiliser. These are usually organic forms of nitrogen (with the exception of urea) that must be broken down into simpler forms before it can be used. Water insoluble nitrogen is referred to as a slow-release nitrogen source and delivers nitrogen at different rates according to the amount and kind of material in its composition.

Organic Fertilisers

The word "organic" applied to fertilisers simply means that the nutrients contained in the product are derived solely from the remains (or a by-product) of a onceliving organism. Urea is a synthetic organic fertiliser, an organic substance manufactured from inorganic materials. Cottonseed meal, blood meal, bone meal, hoof and horn meal and all manures are examples of organic fertilisers. Most of these products packaged as fertilisers will have the fertiliser ratios stated on the package labels.

Some organic materials, particularly composted manures and sludges, are sold as soil conditioners and do not have a nutrient guarantee even though small amounts of nutrients are present. Most are high in one of the three major nutrients and low or zero in the other two although some may be fortified with nitrogen, phosphorus or potash for a higher analysis. Many are low in all three. In general, organic fertilisers release nutrients over a fairly long period. The potential drawback is that they may not release enough of their principal nutrient at the proper time to give the plant what it needs for optimum growth. Because organic fertilisers depend on soil organisms to break them down and release nutrients, most of them are effective only when the soil is moist and soil temperature is warm enough for the soil organisms to be active.

Cottonseed meal is a by-product of cotton manufacturing. As a fertiliser, it is somewhat acidic in reaction. As a fertiliser, it is somewhat acidic in reaction. Formulas vary slightly, but they generally contain 7 percent nitrogen, 3 percent phosphorus, and 2 percent potash. Cottonseed meal is more readily available to plants in warm soils, and there is little danger of burn. For general garden use, apply 2 to 5 pounds per 1000 square feet. Cottonseed meal is frequently used to fertilise acid-loving plants, such as azaleas, camellias and rhododendrons.

Blood meal is dried, powdered blood collected from cattle slaughterhouses. It is a rich source of nitrogen so rich, in fact, that it may do harm if used in excess. The gardener must be careful not to exceed the recommended amount suggested on the label. In addition to nitrogen, blood meal supplies some essential trace elements including iron. Fish emulsion, a well-rounded fertiliser, is a partially decomposed blend of finely pulverized fish. No matter how little is used, the odor is intense but it dissipates within a day or two.

Fish emulsion is high in nitrogen and is a source of several trace elements. In late spring when garden plants have sprouted, an application of fish emulsion followed by deep watering will boost the plant's early growth spurt. Contrary to popular belief, too strong a solution of fish emulsion can burn plants, particularly in containers. Manure is a complete fertiliser, but it is low in the amounts of nutrients it can supply. Manures vary in nutrient content according to the animal source and what the animal has been eating; however, a fertiliser ratio of 1-1-1 is typical. Manures are best used as soil conditioners instead of nutrient suppliers.

Commonly available manures include horse, cow, pig, chicken and sheep. The actual nutrient content varies widely;

the highest concentration of nutrients is found when manures are fresh. As it is aged, leached or composted, nutrient content is reduced. Even though fresh manures have the highest amount of nutrients, most gardeners use composted forms of manure to ensure a smaller amount of salts, thereby reducing the chance of burning plants. Fresh manure should not be used when it will contact tender plant roots. Typical rates of manure applications vary from a moderate 70 pounds per 1000 square feet to as much as 1 ton per 1000 square feet. Reduce rate by 50 percent when using saline irrigation water.

Sewer sludge is a recycled product of municipal sewage treatment plants. Two forms are commonly available: activated and composted. Activated sludge has higher concentrations of nutrients (approximately 6-3-0) than composted sludge and is usually sold in a dry, granular form for use as a general purpose, long-lasting, non-burning fertiliser. Composted sludge is primarily used as a soil amendment and has a lower nutrient content (approximately 1-2-0). There is some question about the long term effect of using sewage sludge products in the garden, particularly around edible crops. Heavy metals, such as cadmium, are sometimes present in the sludge and may build up in the soil.

Possible negative effects vary according to the origin of the sludge as well as with the characteristics of the soil where it is used. In these cases, an analysis of heavy metals in the sludge should be done. Compared to synthetic fertiliser formulations, organic fertilisers contain relatively low concentrations of actual nutrients, but they perform other important functions that the synthetic formulations do not. These functions include increasing organic content of the soil, improving physical structure of the soil, and increasing bacterial and fungal activity, particularly the mycorrhiza fungus, which alone makes other nutrients more available to plants.

Lawn fertilisers

The major reason for buying a lawn fertiliser combined with an herbicide is convenience. It is very convenient to combine everything you need in one application, but it can also be ineffective or even harmful to the trees and shrubs in the lawn area. Sometimes, a fertiliserherbicide combination may actually fertilise the weeds targeted for control. Perhaps an even greater concern with "weed and feed" fertilisers is that the herbicide used may actually be taken up by the trees and shrubs in the lawn area. The result can be a severely damaged or dead specimen tree or shrub hedge.

Nitrogen rates	Formula Pounds per 100 square feet
5-10-10	3.50
6-18-6	2.80
8-12-4	2
12--6-6	1.40
16-16-16	1

Computing the amount of fertiliser needed for a given area is rather tricky at first, but after a few times the logic falls into place and becomes second nature.

Name of fertiliser	%N	%P	%K	Remarks
Blood	10.0	1.5	0.0	A very rapidly available organic fertiliser
Fish scrap	9.0		0.0	Do not confuse with fish emulsives whichare generally quite low in fertiliser content
Guano, bat				Partially decomposed bat manure from caves

Guano, bird		11.0		Partially decomposed bird manure from islands off coast
Kelp or seaweed		0.5		
Meal				
Bone, raw		22.0	0.0	Main value is nitrogen since most of the phosphorus is not soluble
Bone, steamed			0.0	As a result of steaming under pressure, some nitrogen is lost, but more phosphorus is soluble for use by plants
Cocoa shell				Primarily a conditioner for complete fertilisers
Cotton seed	6.0			Generally very acid; useful in alkaline soils
Hoof and horn	14.0	0.0	0.0	The steam-treated and ground material is a rather quickly-available source of nitrogen
Manure				
Cattle	0.5	0.3	0.5	Although manures in general are low in
Chicken	0.9	0.5	0.8	fertiliser, when used in relatively large
Horse	0.6	0.3	0.6	amounts to improve soil structure, damage
Sheep	0.9	0.5	0.8	may occur because of too much salt
Swine	0.6	0.5	0.4	
Mushroom manure (spent)	1.0	1.0	1.0	
Oyster shells	0.2	0.3	0.0	Because of their alkalinity, oyster shells are best suited for raising PH rather than as a fertiliser

Peat (reed or sedge)		0.3	0.3	Best used as a soil conditioner rather than as a fertiliser breaks down too rapidly
Rice hulls (ground)	0.5	0.2	0.5	
Sludge				Examples of activated sludge are Milorganite (Milwaukee, WI), Hu-Acinite (Houston, TX), Chicagrow (Chicago, IL) and Nitroganic (Pasadena, CA)
Sewage				
Sewage, activated (special micro organisms added)	6.0	5.0	0.0	
Tankage				
Cocoa	4.0	1.5		
Garbage	3.0			
Process (leather, hair, wool, felt, feathers etc.)	8.0			
Wood ashes	0.0			Quite alkaline; do not use on high pH soils

Recommendations for fertilising vegetables and annual flowers are usually stated: "Apply 3 to 4 pounds of 5-10-10 fertiliser per 100 square feet of garden space." This is fine as long as the formula, 5-10-10, is used. If the fertiliser to be used has a different formula, perhaps with a higher nitrogen content as indicated by the first number in the formula, the rate of application should be reduced to avoid nitrogen burn. A high phosphorus fertiliser such as 6-18-6 is often recommended for vegetables as a starter food.

Soluble Salts

Nitrogen fertilisers do not burn or damage plants if they are correctly applied. Fertilisers are salts much like our familiar table salt except that they contain various plant nutrients. When a fertiliser (salt) is applied to soil, nearby water begins to very gradually move toward the area where the fertiliser has been applied. Salts and fertiliser begin to diffuse or move away from the place where they were applied. This dilutes the fertiliser and distributes it throughout a much larger area. If tender plant roots are close to the placement of fertiliser, water is drawn from these roots as well as from surrounding soil. The more salt or fertiliser applied, the more water is drawn from nearby roots.

As water is drawn from the roots, plant cells begin to dehydrate and collapse and the plant roots burn or dehydrate to a point beyond recovery. If soil moisture is limited, most of the water drawn toward the salt will come from plant roots and the damage will be severe. Two rules should be kept in mind when applying a fertiliser during hot weather when soil moisture is limited.

Table 1: High salt content Fertilisers

Material	Nutrient level	Relative salinity per weight of nutrient
Ammonium nitrate	33% nitrogen	1.49
Ammonium sulfate	21 % nitrogen	1.63
Potassium nitrate	14% nitrogen	2.67
Natural organic fertiliser	5% nitrogen	0.40
Urea formaldehyde	38% nitrogen	0.13
Urea	45% nitrogen	0.81
Superphosphate	20% phosphorus	0.20
Potassium chloride	60% potash	0.87
Potassium sulfate	50% potash	0.43
Dolomite	30% calcium 20% magnesium	

Gypsum	33% calcium	0.12
Epsom salts	16% magnesium	1.38

Make sure adequate moisture is present after applying fertilisers high in salts. Table 1 lists commonly used garden fertilisers high in salt content or burn potential. The last column is the practical measure of relative salinity. A higher number indicates greater salinity. Soluble salts will accumulate on top of the soil in a container and form a yellow or white crust. A ring of salt deposits may form around the pot at the soil line or around the drainage hole. Salts will also build up on the outside of clay pots.

Soluble salts build up when fertiliser is applied repeatedly without sufficient water to leach or wash the old fertiliser or salts through the soil. It also occurs when water evaporates from the soil, leaving minerals or salts behind. As salts in the soil become more and more concentrated, plants find it harder and harder to take up water. If salts build up to an extremely high level, water can be taken out of the root tips causing them to die.

Soluble salt problems commonly occur on plants grown in containers but rarely occur in the garden. Correct watering is the best way to avoid soluble salt injury by preventing the accumulation of salts. When water is applied, allow some to drain through, and then empty the drip plate. Water equal to one-tenth the volume of the pot should drain through each time water is applied. The pot should not be allowed to sit in water. If drained water is allowed to be absorbed by the soil, the salts that were washed out are taken back into the soil. Salts can be reabsorbed through the drainage hole or directly through a clay pot.

Container plants should be leached every 2 to 4 months. Leach a plant before fertilising to avoid washing away all of the newly-added fertiliser. Leaching is done by pouring a large amount of water on the soil and letting it completely drain. The amount of water used for leaching should equal twice the volume of the pot. A 6-inch pot will hold 10 cups of water, so 20 cups of water are used to leach a plant in a 6-inch pot. Keep the water running through the soil to wash out the salts. If a layer of salts has formed a crust on the soil, remove the salt crust before leaching. Do not remove more than onequarter of an inch of soil. It is best not to add more soil to the top of the pot. If the soluble salt level is extremely high or the pot has no drainage, repot the plant.

The level of salts that will cause injury varies with the type of plant and how it is being grown. A plant grown in the home may be injured by salts at a concentration of 200 ppm. If the same plant is in a greenhouse where light and drainage are good, it will grow with salts 10 times that level or 2,000 ppm. Some nurseries and plant shops leach plants to remove excess salts before the plant is sold. If you are not sure what has been done, leach a newly purchased plant the first time you water it.

Timing and Frequency of Fertiliser Application

Soil type is one factor that dictates the frequency of fertiliser application. Sandy soils require more frequent applications of nitrogen and other nutrients than clay soils. Other factors affecting frequency of application include the type of plants being grown and what is expected of it, the frequency and amount of nitrogen or water applied and the type of fertiliser and its release rate. The plants being grown influence the timing and frequency of fertiliser application since certain plants feed more heavily on some nutrients than others. For example, in the vegetable garden, root crops require less nitrogen than leafy crops. Corn feeds heavily on nitrogen, while most trees and shrubs are generally light feeders. Corn may require nitrogen fertilisation every 4 weeks, while most trees and shrubs perform nicely with a single, well-placed application every year.

Turfgrasses are heavy feeders of nitrogen, requiring at least two applications per year. A general rule of thumb is that nitrogen is for leafy top growth, phosphorus is for root and fruit production, and potassium is for cold hardiness, disease resistance and general durability. Late fertilisation of trees and shrubs can cause new flushes of growth to occur on woody plants that would normally be adjusting themselves for the coming winter. This may delay dormancy in woody plants and cause severe winter dieback of new growth. Table 2 suggests general guides for groups of garden plants. Gardeners should be aware that individual species within these groups vary considerably. After each group of plants, the need for the primary nutrients (nitrogen, phosphorus and potassium) is indicated as high, medium or low.

Table 2: Primary nutrient needs of common landscape plants.

Vegetables	High
Herbs	Medium to low
Lawns	Medium to high
Fruits	Medium
Annual flowers	Medium
Perennial flowers	Medium to low
Deciduous shrubs	Medium to low
Evergreen shrubs	Low
Shade trees	Medium to low

Methods for Fertiliser Application

There are different methods of fertiliser application depending on the formulation and the plant requirements.

Broadcasting: A recommended rate of fertiliser is spread over the growing area and left to filter into the soil, or it is incorporated into the soil with a rototiller or spade. Broadcasting is used over large garden areas, lawn areas or when time or labor are limited.

Banding: Narrow bands of fertiliser are applied in furrows 2 to 3 inches from vegetable garden seeds and 1 to 2 inches deeper than the seeds or plants that are to be planted. Careless placement of the fertiliser band too close to the seeds will burn seedling roots. The best technique is to stretch a string where the seed row is to be planted. With the corner of a hoe, dig a furrow 3 inches deep and 3 inches to one side of and parallel to the string. Spread fertiliser in the furrow and cover with soil. Repeat the banding operation on the other side of the string and then sow seeds underneath the string.

For widely-spaced plants such as tomatoes, fertilisers can be placed in bands 6 inches long for each plant or in a circle around the plant. Place the bands 4 inches from the plant base. If used in the hole itself, place fertiliser at the bottom of the hole. Work it into the soil, placing a layer of soil about 2 inches deep over the fertilised soil; then put the plant in the hole. As the first roots develop, banding is one way to satisfy the need of many plants (especially tomatoes) for phosphorus. When fertiliser is broadcast and worked into soil, much of the phosphorus is locked up by the soil and is not immediately available to the plant. By concentrating the phosphorus in a band, the plant is given what it needs even though much of the phosphorus stays in the soil.

Starter solutions: Another way to satisfy the need for phosphorus when setting out transplants of tomatoes, onions, eggplants, peppers or cabbages is through the use of a liquid fertiliser high in phosphorus as a starter solution. Follow directions on the label.

Side dressing: Dry fertiliser is applied as a side dressing after plants are up and growing. Scatter fertiliser on both sides of the row 6 to 8 inches from the plants. Rake it into the soil and water thoroughly.

Foliar feeding: This is used when a quick growth response is wanted, when micronutrients (such as iron or zinc) are locked in the soil, or when the soil is too cold for the plants to extract or use the fertiliser applied to the soil. Foliar-applied nutrients are absorbed and used by the plant quite rapidly. Absorption begins within minutes after application and is completed within

1 to 2 days with most nutrients. Foliar nutrition can be a supplement to soil nutrition at a critical time for the plant, but it is not a substitute. At transplanting time, an application of phosphorus spray will help establish the young plant in cold soils.

For perennial plants, early spring growth is usually limited by cold soil even when the air is warm. Under such conditions, soil microorganisms are not active enough to convert nutrients into forms available for roots to absorb. Yet, if the nutrients were available, the plants could grow. A nutrient spray for the foliage will immediately provide the needed nutrients to the plants allowing them to begin growth.

COMPOSTING

Making compost is a speeded-up way of imitating nature's cycle of life and death. The end product in the composting process is a dark, loose, partially decomposed, amorphous form of organic matter that reveals no hint of its origin. The only difference between finished compost and forest humus created by nature is time. The actions of various microorganisms and invertebrates convert raw organic matter in a compost pile into finished compost. Maximum decomposition occurs when these organisms are exposed to an appropriate balance of oxygen, moisture, and nutrients. Products of the composting process include finished compost, carbon dioxide, heat, and water.

Aerobic and Anaerobic Composting

Composting can occur under either aerobic (in the presence of oxygen) or anaerobic (without oxygen) conditions. Microorganisms for aerobic composting require oxygen (greater than 5% within the compost pile) while anaerobic microorganisms prefer an absence of oxygen. Anaerobic decomposition in backyard compost piles is generally undesirable due to objectionable odors that result. The most offensive gas resulting from the anaerobic process is hydrogen sulfide (rotten egg smell), while other offensive odors include cadvarine and putrescine. Aerobic decomposition is the preferred composting technique in most backyard composting piles because it smells better and is more efficient. Aerobic composting generates heat up to 140°F or higher, which is hot enough to kill most plant pathogens and weed seed in the pile when properly managed.

Most composting organisms fall into two general groups: microorganisms and invertebrates. Among the microorganisms, aerobic bacteria are the most important in terms of initiating decomposition and heat generation. Bacteria are one-celled colorless organisms that cannot make their own food through photosynthesis. They reproduce by splitting, producing billions of offspring over a relatively short time, although the life span of any particular generation may be only 20-30 minutes. As a group, they can eat almost any type of organic matter, although specific bacterial populations will differ from pile to pile depending upon the makeup of the pile and the decomposition stage. Psychrophilic bacteria are active when a pile is first made, especially in the fall when the weather is cool.

Optimum activity occurs at about 55°F, but these bacteria are still active at 0°F. The bacterial activity creates heat, and sets the stage for the most efficient decomposers, the mesophilic bacteria, who are most active when the temperature of the pile is at 70-100°F. Rapid decomposition of the pile by mesophilic bacteria creates heat. As the temperature of the pile increases, thermophilic bacteria take over from 113°F to 155°F and higher. Most weed seed and organisms causing plant diseases are killed at these temperatures. Unless turned and fed new material, thermophilic bacteria will decompose material rapidly, peaking in 2-3 days.

The pile will then return to the lower temperature range where mesophilic bacteria are active. Other important microorganisms in the composting process include actinomycetes and fungi. Similar to bacteria and fungi, actinomycetes give the compost pile its faintly earthy odor. At the end of the composting process, they may appear as a blue-gray to light green powdery or cobweb-like layer in the outer 4-6 in. of compost. Fungi generally intermingle with the actinomycetes. When the inner pile starts to heat up, most invertebrates are killed or migrate to cooler areas of the pile.

In the cooler areas, nematodes prey upon bacteria, protozoa, and fungal spores. Larger mites and springtails also feed on fungi. The life cycle within the pile continues to become more complex as predaceous mites and pseudoscorpions feed on other mites as well as nematodes. Complex invertebrates like centipedes and ground beetles feed on lower life forms, and decaying plant life in the pile attracts sowbugs, snails, slugs, and earthworms.

Carbon:Nitrogen Ratio

A compost pile that heats up properly and decomposes rapidly depends on a number of environmental factors within the pile. One of the most critical factors is the balance of nitrogen and carbon within the plant and animal wastes added to the pile. Although a number of nutrient ratios, such as nitrogen to phosphorous, are important in the composting process, the most critical is the carbon: nitrogen ratio, commonly expressed as "C:N" or "C/N" ratio. Microorganisms require carbon for energy, while they need nitrogen for the synthesis of protein. Decomposition within the compost pile is optimal when the C:N ratio is about 30:1.

Excess nitrogen in the pile results in the production of ammonia when there is not enough carbon in the pile for microorganisms to synthesize new cellular material. When there is not enough nitrogen to produce proteins for microorganism growth, the composting process slows. While most organic materials in the pile vary in their carbon:nitrogen ratios, they must be balanced to produce the ideal of 30:1. Most dry or woody materials are high in carbon, while green wastes or livestock manures are relatively low (table 3). Because plant materials vary in C:N ratios due to water content and stage of growth, tables of C:N ratios can be used only as guides in mixing materials. Thus, composting is as much an art as it is a science.

Table 3: Carbon:nitrogen ratios for selected organic wastes (by weight).

Material	C:N Ratio
Low C:N Materials	
Grass clippings	12-15:1
Food scraps	15:1
Vegetable waste	11-20:1
Coffee grounds	20:1
Cow manure	18-20:1
Horse manure	25:1
Poultry manure	15:1
High C:N Materials	
Leaves	30-80:1
Straw	40-100:1
Corn stalks	60:1
Paper	170-200:1
Sawdust	200-500:1

Decomposition within the pile can be aided by increasing the surface area of organic material added to the pile. As microbial activity occurs mostly on the surface of waste added to the pile, increasing the surface area will increase the number of microorganisms working on the material.

One of the most efficient ways to increase the surface area of organic waste is to run it through a mechanical shredder or chipper. Shredders and chippers can be large or small, run by electricity, gas, or by hand. Fall leaves can be run through a rotary mower. Loppers and hand shears can be used to reduce the size of hedge trimmings or branches.

Aeration

Turning the compost pile periodically is critical to aerobic composting, as much of the air available to microorganisms within the pile comes from air trapped in it when the pile is turned. The more often the pile is turned, the more air enters it. Some fresh air enters the pile from wind, while air also can be drawn into the pile naturally as warm air from the composting process rises.

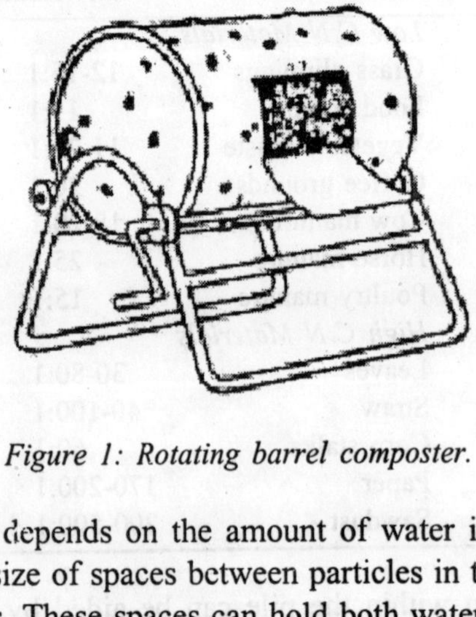

Figure 1: Rotating barrel composter.

Aeration also depends on the amount of water in the pile and the porosity (size of spaces between particles in the pile) of the pile's contents. These spaces can hold both water and air, thus, as water contact increases, possibilities for air contact decrease.

Figure 2: Fin structure in barrel composter.

Mixing coarse dry materials like autumn leaves with wet green materials like grass trimmings will increase the airflow through the pile. In backyard composting projects, a garden fork or pitchfork is one of the best tools to turn or fluff up a pile. Compost piles can also be constructed so that they "breathe" on one or more sides.

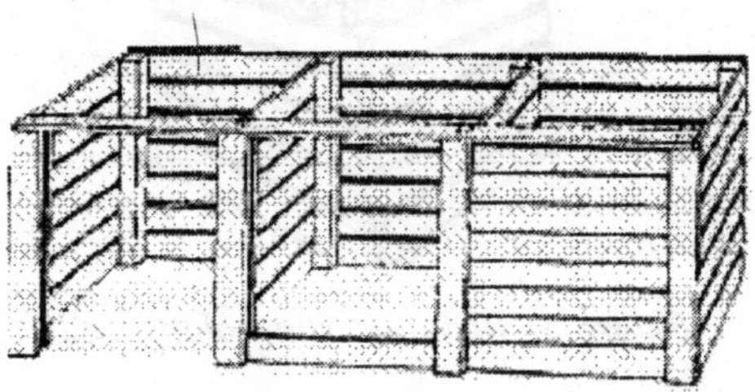

Figure 3: Redwood slat three-bin composting unit.

For optimum aerobic composting, moisture content of the pile should range between 40% and 60%. A higher percentage will increase anaerobic decomposition, while a lower moisture content will slow down the composting process as microorganisms die or become dormant. Squeeze compost in your hand to judge moisture content. If the material feels like a damp sponge and yields only a few drops when squeezed tightly, its moisture content is sufficient.

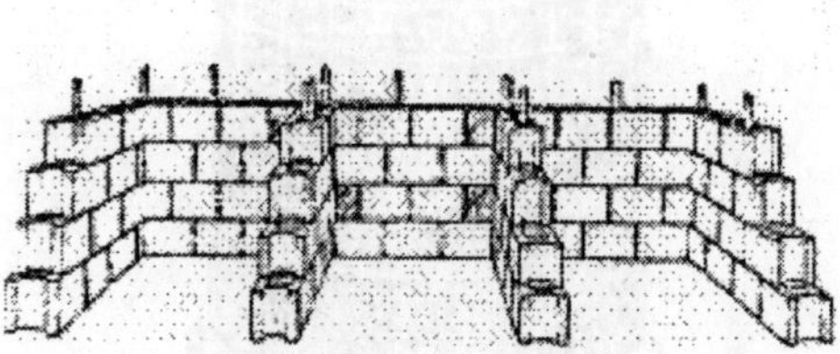

Figure 4: Cement block three-bin composting unit.

Piles that are too wet should be turned to increase air content; piles that are too dry should be turned and sprinkled with a hose.

Making a Compost Pile

Most compost piles aren't aesthetic masterpieces, so locating them in an inconspicuous but convenient location is appropriate. Be sure a source of water is nearby.

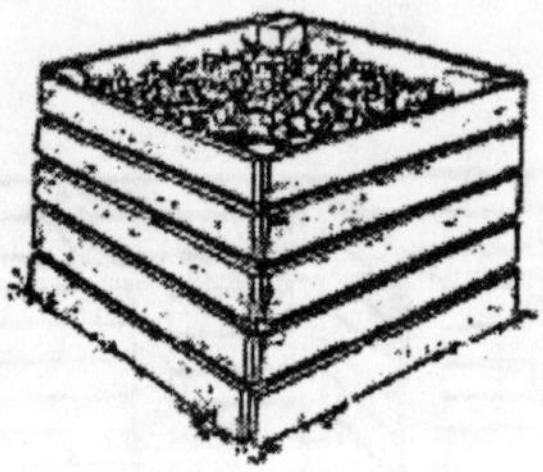

Figure 5: Redwood slat holding unit.

Locating the pile in a shaded area will keep it from drying out. As most compost is used in the garden, easy access to this area is essential.

Minimum dimensions for a pile are 3 × 3 × 3 ft, as smaller piles may not heat up or decompose. The pile should be large enough to hold the heat that it generates, yet be small enough to allow air movement into the center of the pile.

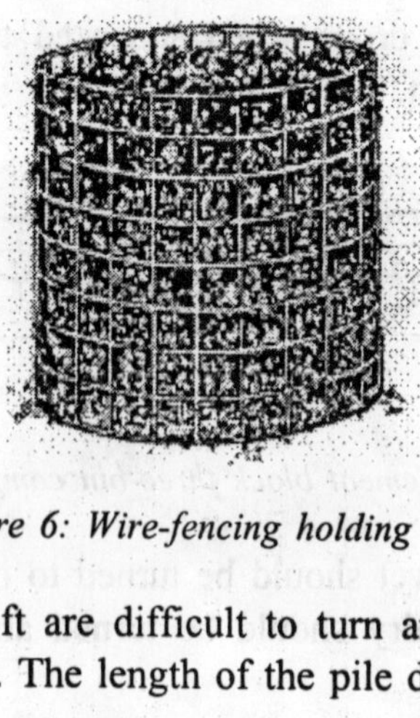

Figure 6: Wire-fencing holding unit.

Piles taller than 5 ft are difficult to turn and may restrict air flow due to weight. The length of the pile depends the amount of material to be composted. Gardeners with large amounts of yard waste should consider constructing two or more compost piles or bins. Two bins are convenient when turning the material, while multiple bins initiated at different times provide for compost at different stages of decomposition.

Structures of a Compost

Making finished compost quickly requires that the structure be adapted to turning the compost frequently. Gardeners who are especially conscious of aesthetics may prefer rotating barrel composters (fig.1).

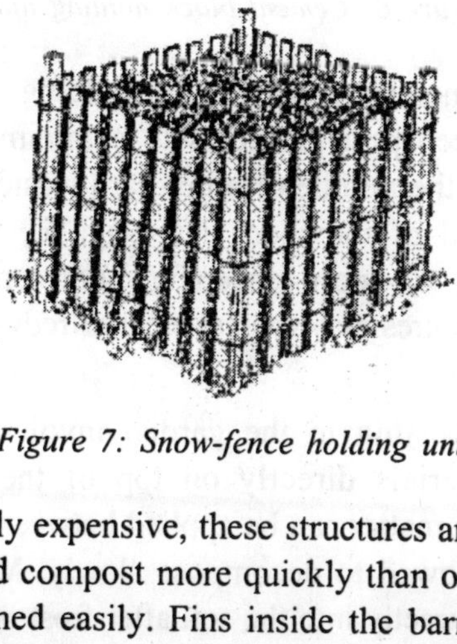

Figure 7: Snow-fence holding unit.

Though relatively expensive, these structures are less messy and produce finished compost more quickly than other structures as they can be turned easily. Fins inside the barrel keep material well aerated (fig.2). Another relatively expensive structure is the three-bin composting unit. These bins can be made of sliding redwood slats (fig.3), a combination of redwood slats and hardware cloth, or cement blocks (fig.4). The hardware cloth structures tend to be better aerated, and compost can be transferred from one bin to another easily.

Heap composting is the cheapest method of composting, but it tends to take up more room and is somewhat messier. Finished compost can be generated quickly if at least two heaps are maintained for turning purposes. "Holding" compost units tend to be slower in terms of creating finished compost,as they tend to be turned less frequently if at all.

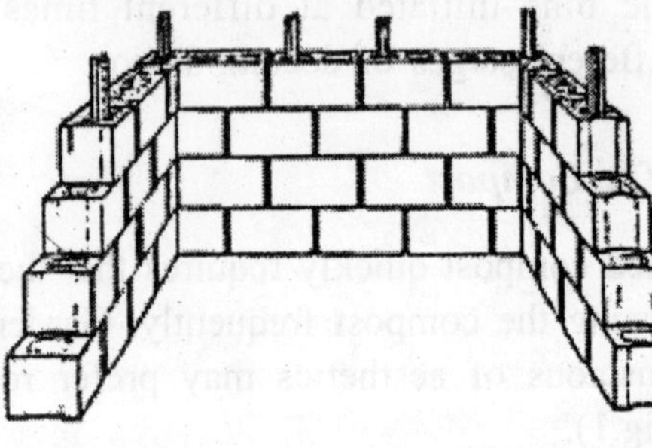

Figure 8: Cement block holding unit.

Materials are generally added to the holding unit as they are generated. To keep pest problems to a minimum, incorporate food wastes into the middle of other yard wastes. Holding units can be composed of redwood slats (fig. 5), wire fencing or chicken wire (fig. 6), snow fence (fig. 7) or cement blocks (fig. 8). All these structures may require steel or redwood corner posts for support.

Sheet composting in the garden involves applying raw composting materials directly on top of the soil in layers. Shredded organic matter can be applied between plants as a type of mulch and allowed to decompose slowly. Material can then be incorporated directly into the soil after frost in the fall. Trench composting involves digging a trench and filling it up with shredded organic waste. The trench is then covered with 6-8 in. of soil. Depending on soil temperature, microorganisms and earthworms will slowly convert these materials to usable organic matter.

Covered trenches are often used as paths between rows of vegetables while the organic matter is decomposing. "Composthole-ing" is similar to trench composting except that it involves smaller areas. A posthole digger or sharpshooter can be used to make holes between plants like tomatoes and these holes can be filled with food scraps and covered with soil. By the following spring, organic matter should be well decomposed.

Techniques for Composting

The most popular composting technique involves alternating layers of organic materials, soil, and fertiliser or manure. Layers can be laid down in either specially constructed bins or piles. A 6-8 in. layer of coarse organic material such as shredded corn cobs or sawdust is first laid down to provide aeration from the bottom. This material will also help absorb leachate that trickles down through the pile. If desired sprinkle a thin layer of fertiliser (1 cup of 12-12-12 or equivalent per 25 square ft of surface rea) over the layer of organic material. Fertilisers provide nitrogen necessary to reduce the high carbon:nitrogen ratios associated with coarse materials like corncobs and sawdust.

Use a complete fertiliser to keep phosphorous and potassium in balance with the nitrogen. A 1-2 in. layer of poultry, rabbit, or livestock manure can be substituted for the fertiliser. Horse manure mixed with bedding materials is an excellent source of nitrogen and microorganisms that start the composting process. Wet grass clippings can also be mixed with drier, coarser materials for a source of nitrogen. Be careful not to layer wet grass clippings in the pile as they tend to mat, and matted wet clippings become sticky masses of slime through anaerobic decomposition, restricting air and water movement in the pile. Continue to layer organic materials, fertiliser or manure, and soil until the pile is 3-5 ft high. Be sure to moisten each layer with water as it's laid down. This is particularly important for drier, coarser materials.

Materials used for Composting

Almost all plant materials can be used in a compost pile. Do not place diseased plants and weed seed in a compost pile, however, as the pile construction may not be optimum for proper heating. In addition, do not put sprigs of perennial lawn grasses like Bermudagrass and root systems of perennial weeds like Johnsongrass and bindweed into your compost pile unless measures have been taken to insure that they do not regrow and reinfest the garden. This can be done by thoroughly drying the material so that they are no longer viable. Whenever possible, select plant materials for composting that will balance high C:N ratio materials with low C:N ratio materials.

Compost piles made of mostly high C:N ratio materials may require extra nitrogen fertiliser or livestock manure to reduce the average C:N ratio for the pile. Animal manures that should not be used include pet (cat and dog) and pig manures that may contain disease organisms that can infect humans. Avoid fatty food wastes like meat and bones, as the odors they generate may attract pests. Avoid wood ashes, as they tend to be very alkaline. Table 4 lists plant and animal wastes to include in the compost pile and those to avoid.

Table 4: Wastes to include in or exclude from the compost pile.

Include:	Exclude:
Leaves both fresh and dried	Weeds with viable seeds
Sawdust	Diseased plantsPig manure
Grass clippings (except Bermuda sprigs)	Pet manure (cat and dog)
Weeds without viable seeds or other propagules	Meat
Vegetable wastes	FatOils and lards
Fruit wastes	Milk and cheese products
Straw	Manmade fabrics such as nylon
Tree and brush trimmings (shredded)	Glass and metals
Hair, natural fabrics	Paper
Egg	Shells

Coffee grounds
Chicken manure
Rabbit manure
Livestock manure (except pig)

Proper, uniform moisture levels for optimum microbial activity can best be maintained when turning the pile. Soggy piles will revert to anaerobic decomposition and need to be fluffed up. Aerobic microbial activity slows in dry piles; thus, when turning dry piles, moisten the compost. Generally, the more compost is turned the faster it decomposes. While frequently turned piles can yield finished compost in as little as 14 days, piles turned less frequently may take as long as a year to produce finished compost.

The time required to develop finished compost also depends on the size of the pile and time of year or air temperature and the size of particles in the pile. When turning, place compost on the outside edges of the pile or bin in the middle of the new pile so that it can heat up in the composting process. The finished compost pile should be half the size of the original pile. Most of the size loss is associated with the production of carbon dioxide in the composting process.

VERMICOMPOSTING

Vermicompost contains not only worm castings, but also bedding materials and organic wastes at various stages of decomposition. It also contains worms at various stages of development and other microorganisms associated with the composting processing. Earthworms are some of the largest organisms found in compost. Tunneling through garden soil, garden earthworms can consume as much as their own body weight in soil each day. Castings they leave behind are five to eleven times richer in nitrogen, phosphorous, and potassium than the soil they eat.

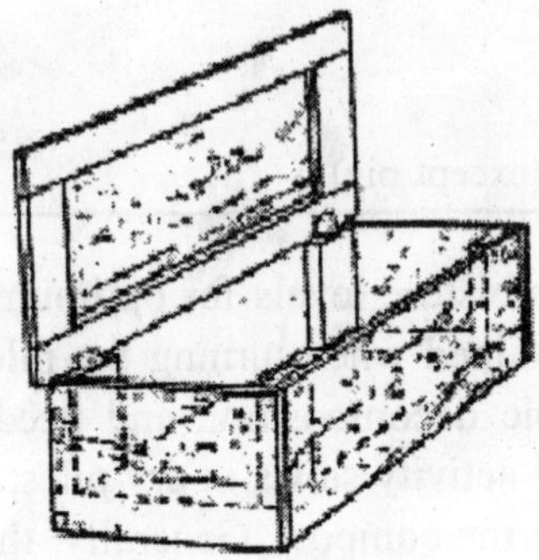

Figure 9: Worm Compost Box

The garden earthworm and the compost earthworm, however, are radically different. Although garden or field earthworms (Allolobophora caliginosa) will penetrate the bottom of a compost pile, they prefer regular soil and will not tolerate active compost with warmer temperatures. Red worms or red wigglers (Lumbricus rubellus) and brandling worms (Eisenia foetida), however, prefer cooler compost or manure piles and will die off if the compost is too hot.

Specialized bins or boxes (fig. 9) can accommodate worm compost or vermicompost. Red worms and brandling worms prefer temperatures of 50-70°F. Shallow ventilated wooden or plastic boxes with lids are first filled with moist bedding (50-75% water, by weight) made of peat moss, sawdust, or shredded newspaper. Worms are then added and allowed to crawl down through the bedding. For the beginning vermicomposter about 100 worms will quickly multiply and populate a small bin such as can be found at any grocery store. Pockets of food waste can be buried in the bedding, and if the bed is kept moist, the worms will turn the food wastes and bedding into a high-quality soil amendment.

The worms will normally multiply and become efficient producers of castings. To separate the worms from the castings expose the contents of the bin to light and the worms will hide from the light and the castings can be scraped from the top of the container. When exposed on the surface the worms will again crawl away from the light and more castings can be removed. Con-tinue this process until all that is left at the bottom of the container is a mass of intertwined worms. These can be used to start additional boxes if more are desired or they can be put in the outdoor compost pile.

Uses of Compost

Finished compost has many uses in the home garden. As it is created from a mixture of plant and animal wastes, it contains a broad base of both macro and micro fertiliser nutrients. Most of these nutrients are bound in an organic form and are made available to plants slowly throughout the growing season. The relative quantity of any one nutrient found in compost is quite low; thus, the major benefit of compost lies in its soil conditioning qualities. As a rich source of organic matter, it improves the water holding capacity of sandy soils and their ability to retain nutrients and release them to plant roots as needed.

In heavier clay soils, compost binds clay particles together to form aggregates, improving both water drainage and the ability of plant roots to penetrate the soil. Conditioning soil surface layers with compost will also improve water penetration and decrease soil erosion and soil crusting. As a soil conditioner, compost contributes organic acids that weather mineral portions of the soil, making nutrients more available for plant uptake. Compost also improves the buffering capacity of the soil, helping stabilize soil pH levels. Because earthworms are attracted to organic matter, earthworm populations increase and make more nutrients available from deeper in the soil.

Compost is most frequently applied to garden soils in the spring before planting. Compost should be liberally applied and turned under to a depth of 4-6 in. Gardeners with limited compost should incorporate it only below the seed row. Handfuls of compost can also be incorporated in planting holes for transplants. Compost also makes an excellent top dressing for lawns, trees, shrubs, and perennial flowers. Mulches around vegetables and other plants help reduce water evaporation from the soil and cut down on annual weed growth. as well as serving as a slow release source of nutrients. Compost will also cool the soil and buffer soil water fluctuations.

Yard and food waste make up a major component of solid waste in most municipalities throughout the United States. Although much of this organic waste can be recycled in the backyard using traditional aerobic backyard composting techniques, these techniques are not appropriate for apartment dwellers and are often inconvenient, particularly during bad weather in the winter. Vermicomposting, or composting with earthworms, is an excellent technique for recycling food waste in the apartment as well as composting yard wastes in the backyard. Worm bins located near a hot water heater in the garage during the winter will save many a trip through the snow to the backyard compost bin. Letting worms recycle your food waste also saves your back, because you don't have to turn over the compost to keep it aerated.

The most common types of earthworms used for vermicomposting are brandling worms (Eisenia foetida) and redworms or red wigglers (Lumbricus rubellus). Often found in aged manure piles, they generally have alternating red and buff-colored stripes. They are not to be confused with the common garden or field earthworm (Allolobophora caliginosa and other species). Although the garden earthworm occasionally feeds on the bottom of a compost pile, they prefer ordinary soil. An acre of land can have as many as 500,000 earthworms, which can recycle as much as 5 tons of soil or more per year.

Redworms and brandling worms, however, prefer the compost or manure environment. Passing through the gut of the earthworm, recycled organic wastes are excreted as castings, or worm manure, an organic material rich in nutrients that looks like fine-textured soil. Earthworm castings in the home garden often contain 5 to 11 times more nitrogen, phosphorous, and potassium as the surrounding soil. Secretions in the intestinal tracts of earthworms, along with soil passing through the earthworms, make nutrients more concentrated and available for plant uptake, including micronutrients. Redworms in vermicompost act in a similar fashion, breaking down food wastes and other organic residues into nutrient-rich compost. Nutrients in vermicompost are often much higher than traditional garden compost.

Finished vermicompost should have a rich, earthy smell if properly processed by worms. Vermicompost can be used in potting soil mixes for house plants and as a top dressing for lawns. Screened vermicompost combined with potting soil mixes make an excellent medium for starting young seedlings. Vermicompost also makes an excellent mulch and soil conditioner for the home garden.

Construction of a Worm Bin

Bins can be made of wood or plastic, or from recycled containers like old bathtubs, barrels, or trunks. They also can be located inside or outside, depending on your preferences and circumstances. As red wigglers tend to be surface feeders, bins should be no more than 8 to 12 inches deep. Bedding and food wastes tend to pack down in deeper bins, forcing air out. Resulting anaerobic conditions can cause foul odors and death of the worms.

The length and width of the bin will depend on whether it is to be stationary or portable. It also depends on the amount of food waste your family produces each week. A good rule of thumb is to provide one square foot of surface area per pound of waste in your bin. Wooden bins have the advantage that they're more absorbent and provide better insulation. Do not use redwood or other highly aromatic woods that may kill the worms. Plastic tends to keep the compost too moist. Plastic, however, tends to be less messy and easier to maintain. Be sure containers are well cleaned and have never stored pesticides or other chemicals. Drilling air/ drainage holes (1/4- to 1/2-inch diameter) in the bottom and sides of the bin will ensure good water drainage and air circulation.

Place the bin on bricks or wooden blocks in a tray to catch excess water that drains from the bin. The resulting compost tea can be used as a liquid fertiliser around the home landscape. Each bin should have a cover to conserve moisture and exclude light. Worms prefer darkness. Bins can be covered with a straw mulch or moist burlap to ensure darkness while providing good air ventilation. Outside bins may require a lid to exclude scavengers and other unwanted pests. Outdoor bins should be insulated from the cold to protect the worms. One option is to dig a rectangular hole 12 inches deep and line the sides with wooden planks. The bottomless box can then be filled with appropriate bedding material, food wastes, and worms.

Food wastes can be continually added as they accumulate. The pile should be kept damp and dark for optimum worm activity. During the winter, soil can be piled against the edges of the bin and straw placed on top to protect the worms from cold weather. Do not add food waste to outdoor bins during the winter because this could expose the worms to freezing weather.

Bedding for bins can be made from shredded newspapers (non-glossy), computer paper, or cardboard; shredded leaves, straw, hay, or dead plants; sawdust; peat moss; or compost or aged (or composted) manure. Peat moss should be soaked for 24 hours in water, then lightly wrung out to ensure it is sufficiently moist. Grass clip-pings should be allowed to age before use because they may decompose too quickly, causing the compost to heat up. Bedding materials high in cellulose are best because they help aerate the bin so the worms can breathe. Varying the bedding material provides a richer source of nutrients.

Some soil or sand can be added to help provide grit for the worms digestive systems. Allow the bedding material to set for several days to make sure it doesn't heat up (and allow to cool before adding worms). The bedding material should be thoroughly moistened (about the consistency of a damp sponge) before adding the worms. Fill the bin three-quarters full of moist bedding, lifting it gently afterwards to create air space for the worms to breathe and to control odors.

Under optimum conditions, redworms can eat their own weight in food scraps and bedding in one day. On the average, however, it takes approximately 2 pounds of earthworms (approximately 2,000 breeders) to recycle a pound of food waste in 24 hours. The same quantity of worms requires about 4 cubic feet of bin to process the food waste and bedding (1 cubic foot of worm bin/ 500 worms).

Earthworms eat all kinds of food and yard wastes, including coffee grounds, tea bags, vegetable and fruit waste, pulverized egg shells, grass clippings, manure, and sewage sludge. Avoid bones, dairy products, and meats that may attract pests, and garlic, onions, and spicy foods. Limited amounts of citrus can be added, but too much can make the compost too acidic. The compost should be kept at a pH of 6.5 if possible, with upper and lower limits at 7.0 and 6.0, respectively. Overly acidic compost can be corrected by adding crushed eggshells.

Avoid adding chemicals (including insecticides), metals, plastics, glass, soaps, pet manures, and oleanders or other poisonous plants, or plants sprayed with insecticides to the worm bin. Food wastes should be added to the bin by pulling back the bedding material and burying it. Be sure to cover it well to avoid attracting flies and other pests. Successive loads of waste should be buried at different locations in the bin to keep the food wastes from accumulating. Grinding or blending the food waste in a food processor speeds the composting time considerably.

Redworms can survive a wide range of temperatures (40-80°F), but they reproduce and process food waste at an optimum bedding temperature range of 55-77°F. The worms should never be allowed to freeze. Bins kept outside may have to be insulated with straw in the winter to keep the worms from freezing. Portable bins can be kept by a hot water heater in the garage during the winter to keep them warm. The bin contents should be kept moist but not soaked. Do not allow rainfall to run off a roof into the bin. This could cause the worms to drown. A straw covering may be needed in exposed sites to keep the bin from drying out during hot summer weather.

Food scraps can be continually added to the bin for up to 2 to 3 months, or until you notice the bedding material disappear. When the bedding disappears, harvest the worms and finished compost, then refill the bins with new bedding material. Overloading the bin with food wastes can result in foul odors. If you notice these odors, stop adding the waste until the worms have a chance to catch up. Overly moist food waste and bedding also cause odors. To relieve this problem, fluff up the bedding to add air and check the drainage holes. As a general rule of thumb, keep the bedding material moist, but never soggy.

Make sure the food waste is buried properly in the bedding. Exposed food wastes can attract fruit flies, house flies, and other pests. Keeping the bin covered with straw or moist burlap also deters these pests. Garden centipedes can be a problem in the worm bin, especially outside. These predators should be destroyed. Overly wet beds also can attract the earthworm mite, which may cause the worms to stop eating.

Separation of Worms from Compost

There are three basic ways to separate the worms from the finished compost. One way involves moving the finished compost and worms over to one side of the bin and adding new bedding material and food waste to the other side. Worms in the finished compost should move over to the new bedding with the fresh food waste. The finished compost can then be removed. A second way to remove the worms is to build a small harvester frame of 2 × 4s with a 3/16-inch mesh bottom. Place the worm compost on the frame and sift the worms out. Larger pieces of compost can be returned to a new batch of bedding and worms.

The compost also can be placed in small piles on a tarp in the sun (or under bright lights inside). Because worms don't like light, they will wiggle to the bottoms of the piles. After waiting 10 minutes, remove the upper inch or more of finished compost from each pile until you run into the worms. Allow the worms to again wiggle to the bottom of the pile and repeat the process. Combine what's left of the small piles into one big pile and again repeat the process. You should eventually end up with a pile of finished compost and a ball of worms. The worms can be added back to a new bin of bedding and food waste. Larger worms also can be used as bait for fishing.

SOIL TESTS

Having the soil tested is an excellent way to know the fertility level of soil and ho much fertiliser to add.

Major Plant Nutrients

The major plant nutrients are nitrogen (N), phosphorus (P), and potassium (K). Commercial fertiliser labels usually show, in percentages, how much actual nitrogen (N), phosphorus as available phosphoric acid (P_2O_5), and potassium as potassium oxide (K_2O) they contain. Fertiliser recommendations are often given in terms of N, P, and K. To convert a recommendation for P into the P_2O_5 form, which is most common, multiple the P value by 2.29. Similarly, multiple the K value by 1.2 to obtain the recommended amount with K_2O. Thus, a 20-20-20 (N-P-K) recommendation would mean the fertiliser contained 20 units of nitrogen, 45.8 units of phosphorus, and 24 units of potassium.

Fertiliser recommendations are often made in terms of pounds per acre. However, most gardeners plant areas much smaller than an acre.

Vegetable	Nitrogen (N) (lb)	Phosphorus (P_2O_5) (lb)
Cabbage, broccoli, cauliflower	5.0	2.0
Carrots	2.8	1.4
Chile	1.8	1.0
Corn (sweet)	4.5	1.8
Cucumbers	2.3	1.8
Green beans	0.5	1.4
Pinto beans	0.5	1.0
Lettuce, onions	4.5	2.3
Potatoes	4.5	4.1
Tomatoes	3.7	3.4
Melons common	1.8	1.8

Commercial fertiliser comes in many different formulations that supply different amcunt of nutrients. Therefore, calculations are necessary to determine how much product is required to supply needed nutrients. The gardener will need to apply 3.91 lb of urea per 1,000 sq ft for chile. Too much nitrogen can burn plants. If you are applying a mixed fertiliser, apply amounts according to the nitrogen requirement to avoid applying too much nitrogen. The N requirement is 3.7 lb per 1,000 sq ft. You have decided to use 18-12-6 fertiliser the proportion needed for one unit of N is 5.6. Multiply the fertiliser recommendation (3.7) by the proportion needed (5.7).

For vegetable gardening, do not use fertilisers that have herbicides or pesticides mixed in them, as plants are easily burned when these are applied to the garden. Some gardeners may want to use organic fertilisers. Phosphorus becomes available to plants gradually, so it can be applied all at one time before planting. Nitrogen applications, however, should be split. Provide ni-trogen in two or three applications, which allows plants to feed more evenly and decreases the possibility of salt burn. As a general rule, apply no more than 1 lb of nitrogen at one time for 1,000 sq ft of garden space.

When the total nitrogen requirement is more than 1 lb, supply it in two or more applications. Apply about one-third of the required nitrogen before planting, apply one-third when plants reach four to five inches tall, and apply the remaining one-third about a month later. This system of split applications is also good for applying fertilisers.

Planting Vegetables

Home vegetable gardeners tend to have a potpourri of vegetables in a small area, so fertilising different crops with differing amounts can be tedious. However, if the vegetables are grouped according to their fertiliser needs, then only two or three different fertiliser amounts have to be applied.

Fertilisation in Small Garden Areas

If the garden area is small, a compromise fertiliser application can be made. The application will not provide the correct amount of nutrients for each crop, but it should provide enough nutrients for adequate growth and development. For most gardens and most vegetables, 3 lb of nitrogen and 2 lb of phosphoric acid per 1,000 sq ft is suitable. A complete fertiliser is one that includes nitrogen, phosphorus and potassium, and it should supply the plants' nutrition requirements from planting to harvest. Unfortunately, the recommended rates of commercial fertilisers are often given on the per-acre basis.

CHAPTER 3

NITROGEN FERTILISATION

Nitrogen (N) is the essential plant element that most frequently limits irrigated crop production. Commercial N fertilisers are a cost-effective means of supplementing soil supplied N for plant growth and are necessary for sustaining high crop yields. However, it has been documented that improper or excessive use of N fertiliser can lead to nitrate pollution of ground or surface water. Both urban and rural fertiliser applicators can minimize this problem by implementing Best Management Practices (BMPs) for fertiliser use.

Nitrate (NO_3) is a naturally occurring form of N that is highly soluble in water and may cause health problems if ingested in large amounts. A number of sources of NO_3 exist, including manure, septic and municipal effluent, decomposing organic matter, soil organic matter, and N fertiliser. High NO_3 levels in drinking water can cause methemoglobinemia, or "blue baby syndrome," a condition primarily seen in very young infants and farm animals. Although reports of methemoglobinemia are extremely rare, the U.S. EPA has established a safe drinking water standard of 10 ppm NO_3-N for community drinking water supplies. Managing the amount, form, placement, and timing of N application is the most practical and acceptable approach to minimizing ground and surface water contamination resulting from fertiliser use.

Nitrogen fertilisation is also an environmental issue. If fertiliser N is applied in excessive amounts or before heavy precipitation, it may give rise to increased leaching and denitrification losses. Some fertilisers, e.g. farmyard manure, may lose large amounts of N through gaseous emissions of ammonia during and after application, depending on spreading technique and weather conditions.

NITROGEN CYCLE

Fertiliser use efficiency is at, or near, an all-time high in U.S. agriculture. Total fertiliser use has remained relatively level since the mid 1980s, while crop yields continue to increase. In spite of this, it is estimated that about 25% of all preplant N applied to corn is lost through leaching or denitrification. This loss is estimated to be even higher during wet years. Leaching is the major problem on coarse-textured soils, while denitrification is the primary pathway on poorly drained clay soils. While a certain amount of loss is unavoidable, producers can gain economic and environmental benefits by minimizing losses and maximizing crop uptake.

To fully understand the transformation and movement of N in the environment, some knowledge of the N cycle is needed. Nitrogen in the soil is commonly found in the form of organic N in the soil humus, ammonium (NH_4), nitrate (NO_3), or in a gaseous form (NH_3, N_2, N_2). Nitrogen in soil organic matter may be converted to the NH_4 form by a biological process called mineralization. The NH_4 form is converted to NO_3 by another biological process called nitrification (Figure 1). Fertiliser N, whether organic or inorganic, is biologically transformed to NO_3, which is highly leachable. The speed of this transformation is determined by soil temperature and moisture, but will eventually occur in any well-drained agricultural soil. Plants will absorb and utilize both NH_4 and NO_3. Therefore, producers need to match N applications to crop uptake patterns to minimize NO_3 leaching and maximize efficiency.

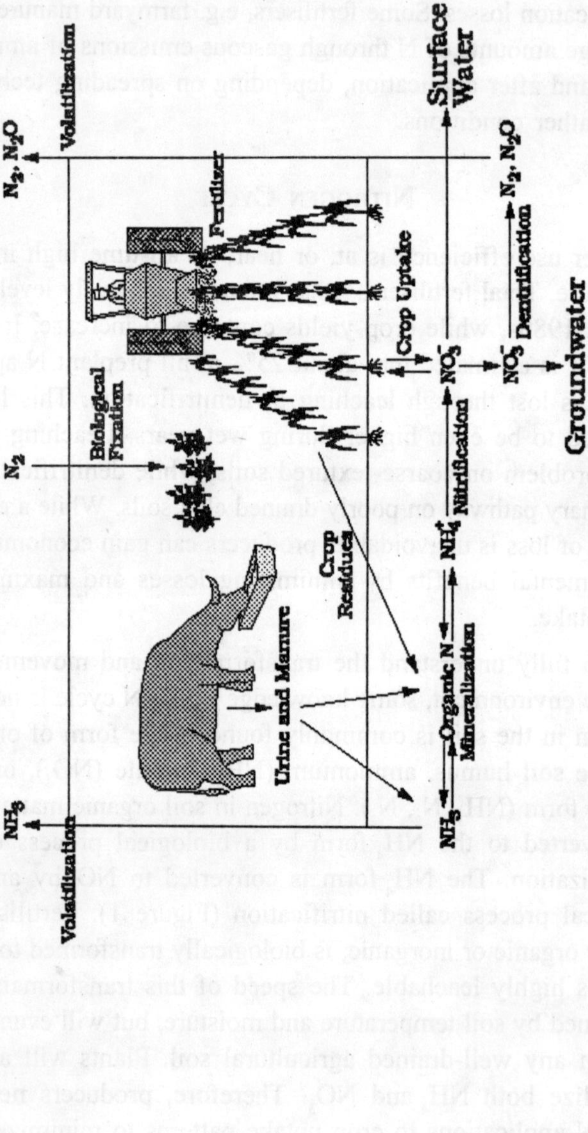

Figure 1: The nitrogen cycle

LEACHING HAZARD

Leaching potential of a given site depends upon soil properties, management, irrigation, and climatic factors. Depth to groundwater and the overlying geologic material determine the contamination potential of an aquifer. Due to the site-specific nature of these properties, applicators must determine the relative leaching hazard of each application site in order to select the appropriate BMPs (Table 1). Leaching hazard can be ranked as severe, moderate, or slight by simultaneously considering soil characteristics, irrigation method, and aquifer vulnerability. Those operators with sites that have a severe leaching potential and a shallow water table (<25 feet deep) must select appropriate BMPs to decrease leaching hazard.

Table 1: Potential leaching hazard as predicted by soil type1, aquifer depth, and irrigation method

	Aquifer Depth and Condition		
Irrigation method	Shallow or contaminated	Deep or uncontaminated	Aquifer properties unknown
Flood or Conventional furrow with efficiency <60%	Coarse soil: Serve Fine soil soil: *Moderate*	Coarse soil: *Moderate* Fine soil: *Slight*	Coarse soil: *Severe* Fine soil: *Moderate*
Sprinkler or surge furrow with efficiency > 60%	Coarse soil: *Moderate* Fine soil: *Slight*	Coarse soil: *Moderate* Fine soil: *Slight*	Coarse soil: *Moderate* Fine soil: *Slight*
Dryland or drip irrigation	Coarse soil: *Moderate* Fine soil: *Slight*	Coarse soil: *Slight* Fine soil: *Slight*	Coarse soil: *Moderate* Fine soil: *Slight*

Operators working under moderate leaching conditions should assess what particular practices may cause future groundwater contamination and make the necessary changes to prevent any groundwater quality problems. Operators with uncontaminated groundwater and slight leaching hazard should continue observing good management practices. Table 1 can help operators evaluate the potential leaching hazard of their particular sites. Information on the depth to the water table and water quality can be obtained through several sources, if not currently known. All rural well owners are strongly encouraged to periodically have a water sample analyzed by a qualified laboratory.

Experienced local well drillers also have knowledge of local groundwater conditions. Agencies such as the Soil Conservation Service, Cooperative Extension, and others can frequently provide information to help you evaluate groundwater vulnerability at your site. In some cases, soil type and a shallow water table combine to create a very high leaching potential. For example, if the soil at a given site is course textured, and depth to the water table is less than 10 feet, it is strongly recommended that shallow-rooted crops not be grown under conventional furrow irrigation. Deeper rooted crops and higher efficiency irrigation methods are necessary for these conditions.

While soil, climatic, and geologic characteristics of the site strongly influence leaching potential, management practices finally determine the amount and extent of N leaching. Proper nutrient management includes:

— accounting for crop N needs

— applying appropriate inputs as determined by N budget

— applying N when and where it can be used most efficiently by the crop.

This will assure residual soil NO_3 is minimized. The following management practices will help producers and fertiliser applicators maximize economic returns from fertiliser dollars while protecting water quality.

Soil Testing

Soil testing is a very important BMP for determining plant nutrient needs. Yearly sampling of each field is necessary to make accurate N fertiliser recommendations. The key to good soil test results is proper sampling protocol. Each sample should contain 12 to 20 cores of soil from a reasonably uniform area of approximately 40 acres. Large fields should be broken into sampling units based upon crop, yield, and fertiliser histories. Deep soil sampling for residual NO_3 is requisite to precise fertiliser recommendations and provides producers season-end information regarding crop N use and N remaining for next year's crop. Sampling to a minimum depth of 2 to 3 feet is recommended for all soil types.

Setting realistic yield goals is also a very important BMP. Fertiliser N recommendations are based upon a yield goal submitted by producers with their soil samples. While farmers tend to be optimistic, overestimating yield goals results in excess N applications, leading to loss of farm income and potential groundwater contamination. Applying enough fertiliser for a 200 bu/A corn crop, when other conditions such as limited irrigation water will only allow a 150 bu/A yield, can result in 60-70 lb/A of excess N being applied. Rather than project a yield goal, it is recommended that producers establish a yield expectation based upon historical yield averages. Yield expectations must be established on a field-byfield basis.

The five most recent yield averages for each field should represent an obtainable yield. If a recent crop has been lost to hail or other disaster, that year's yield should be omitted from the average. Soil organic matter, irrigation water, manure, and previous legume crops all contribute N to the growing crop. The N contribution from these sources must be credited in order to make accurate fertiliser recommendations. Table 2 suggests average credits from various sources of N. Legume crops can be a very significant source of plant available N due to bacterial

N_2 fixation in root nodules. Plowing down a full stand of alfalfa will release as much as 100 pounds of N per acre in the first year after plowdown. The amount of N credit given for legumes depends upon the crop, stand, and degree of nodulation. A minimum of 30 lb N/acre should be credited in the first year after any legume crop.

Irrigation water containing nitrate can supply N to the crop since it is applied and taken up as the crop is actively growing. Water tests for NO_3-N should be taken annually to accurately calculate this credit. Multiply ppm NO_3-N by 2.7 lb/AF times the amount of effective water applied (in feet) to the crop to determine pounds of N per acre applied in the irrigation water. Inexpensive quick tests are available for on-farm water testing. If a water sample is taken for laboratory analysis, it should be kept cool, but not frozen, until it gets to the lab.

Table 2: Nitrogen credits for crop requirements

N Source	N Credit
Soil organic matter	30.0 lb N/%OM
Residual soil nitrate	3.6 lb
N/ppm NO_3-N Manure	10.0 lb
N/ton dry manure	2.7 lb N/AF x ppm NO3-N
Irrigation water	
Previous alfalfa/sweet clover	50.0 lb N/A
Other previous legume crop	30.0 lb N/A

Manure is an excellent plant nutrient source, but excessive manure applications can result in water pollution from soil buildup of N and P. For best results, manure should be analyzed prior to application to determine nutrient content. Subsequent manure application rates should be determined by factoring the previous application to avoid excess loading. In general, about 50% of total N in manure is available the first year after application, 25% in the second year, and 12.5% in the third year. In the absence of an accurate manure analysis, operators should credit 10 lb N/ton of dry manure applied the first year after application. Sewage sludge is another valuable source of plant nutrients that must be properly used to avoid environmental problems. One ton of dry sludge may contain up to 100 pounds of total N, 120 pounds P_2O_5, and 10 pounds K_2O at a fertiliser value of $30 to $60 per ton.

Timing and Optimal Fertiliser Placement

Proper timing and optimal fertiliser placement can greatly enhance plant uptake of N. Subsurface applied or incorporated fertiliser is much less subject to surface losses than surface broadcast fertiliser. Band applied fertiliser can be placed in closer proximity to plant roots. All surface applied fertilisers should be incorporated to reduce runoff and volatilization. In furrow irrigated cropland, alternate furrow irrigation used with ridge banded fertiliser can significantly reduce downward movement of N.

Application of N through high efficiency irrigation systems such as center pivot or surge systems at periods of maximum crop uptake may increase N efficiency. It is not recommended that N be applied through low efficiency furrow or flood systems due to runoff and deep percolation losses. In higher efficiency surface systems such as surge irrigation, tailwater recovery should be employed to capture and recycle nutrients. Surface application of fertiliser or manure on frozen or snow-covered sloping fields should also be avoided. Fertiliser applications should be timed to coincide as closely as possible to the period of maximum crop uptake.

Fall applied N fertiliser has been shown to cause groundwater degradation in areas of high fall and winter precipitation. It should be avoided on spring planted crops in situations with severe potential leaching hazard (Table 1). There may be economic and management benefits to applying N in the fall, but the environmental risks make this a poor choice on coarse-textured soils or in situations where preplant irrigation is necessary. Partial application of N in the spring, followed by sidedress application, improves crop N uptake efficiency and reduces N available for leaching.

Waiting until the crop is well established before applying large amounts of N reduces early season losses and allows producers to more accurately determine the crop yield potential. Poor stands, poor weed control, and below average precipitation are good reasons to adjust N rates downward at sidedress time. Conversely, exceptional conditions warrant increased N at sidedress. This type of managerial flexibility offers producers economic benefits and helps maintain water quality.

Nitrate forms of N fertiliser are readily available to crops, but are subject to leaching losses. Nitrate forms should not be applied in large amounts when the leaching hazard is moderate to high. Ammonium N forms, such as urea or anhydrous ammonia, are preferred in these situations because they are not subject to immediate leaching. However, under warm, moist soil conditions, transformation of NH_4 to NO_3 occurs rapidly. Other more slowly available N sources are commercially available and should be used where they are economically feasible. Nitrification inhibitors can be used to delay the conversion of NH_4 to NO_3 under certain conditions. Farmers should consider using nitrification inhibitors when it is not feasible to use split applications or other management techniques on leachable soils. Nitrification inhibitors seldom produce a direct economic return to farmers and should not be used as a substitute for following other BMPs, but they can reduce leaching under certain situations.

Plant Analysis

Plant analysis during the growing season is another practice to help assess nutrient sufficiency in the growing plant. While nutrient deficiencies are many times visibly apparent, excess nutrient levels can only be determined by plant tissue analysis. This technology offers producers the ability to apply lower rates of N preplant, and to monitor and adjust plant nutrient status throughout the growing season. Plant analysis, when properly used, offers producers insurance that careful N management will not negatively affect the bottom line.

Irrigation Management

Under dryland agriculture, the most significant water quality impacts are usually from soil erosion and subsequent nutrient enrichment of surface water. Irrigated crop production has the greatest potential to cause groundwater contamination due to the large volume of water applied. Nutrients leached below the crop root zone are lost to the system and may become pollutants. Crop root zone depth will vary by soil type, irrigation method, and management (Table 3). Therefore, conscientious management of irrigation water is critical to proper N management. Increasing irrigation efficiency and uniformity reduces the amount of water drained through the soil, and decreases the amount of NO_3 and other contaminants leached.

Table 3: Approximate rooting depths for selected crops under furrow irrigation Crop Root Depth at Maturity (ft)

Corn	3-6
Small grains	3-5
Onions	1-2
Sugarbeet	5-8
Alfalfa	8-15
Dry beans	2-3

A number of technologies allow producers to apply water uniformly without excessive waste, and help to determine the optimum timing and amount of water to be applied. Among these are irrigation systems such as sprinkler, low-energy precision application, surge, and drip. Delivery systems such as lined ditches and gated pipe, as well as reuse systems such as tailwater recovery ponds, can greatly enhance overall efficiency. Shortening irrigation run lengths on coarse-textured soils can decrease deep percolation and leaching. Laser leveling of fields, irrigation scheduling according to soil water depletion and plant needs, and conservation tillage methods can also significantly decrease irrigation requirements. Proper irrigation scheduling, coupled with efficient irrigation systems, are among the most important BMPs to reduce nitrate leaching.

N Management Tools

Although proper N rates and good irrigation management are the most critical components of N management, there are other tools that should also be considered. Proper calibration and maintenance of fertiliser equipment is essential to get uniform distribution of fertiliser at the correct rate. Crop rotation can be beneficial by minimizing total fertiliser and pesticide needs. Often, yield improvement and economic benefits are achieved through a good rotation plan that takes advantage of better pest control, soil tilth, and N fixation by legumes. Deep-rooted crops can be used to scavenge N left in the subsoil by shallow-rooted crops. Cover crops are beneficial in preventing wind and water erosion, and they can use residual N in the soil profile. Finally, computer-assisted decision aides such as the Nitrate Leaching and Economic Analysis Package model can help producers make wise choices and avoid unnecessary water quality degradation.

EFFECTS OF NITROGEN FERTILISATION

Nitrogen deposition from the atmosphere has strongly increased in Europe and other industrial regions world-wide since the middle of the 20th century. This is due to increased emissions of ammonia (NH_3) and nitrogen oxides (NO_x). Ammonia is volatilized primarily in regions with intensive agricultural systems. The massive use of nitrogen fertilisation and, more importantly, high livestock densities result in enhanced emission of NH_3–N and thus in higher deposition rates of NH_4–N, predominantly as $(NH_4)_2SO_4$.

Nitrogen oxides originate mainly as a side product when fossil fuel is burnt in automobile engines and in industrial processes. This problem has been considerably reduced but not eliminated through the widespread use of "catalysers", which convert the nitrogen oxides back into nitrogen and oxygen. Nitrogen oxides are deposited primarily in the form of NO and NO_2.

These anthropogenic processes have increased the input of biologically reactive forms of N into the global N cycle. Compared with natural N fixation, these human-related activities have more than doubled the input of N into the global N cycle. Because of short- and long-range transport of these nitrogenous compounds, atmospheric nitrogen deposition has increased in many ecosystems across the world and has therefore become a major concern for the health of forests. Currently, the areas with high atmospheric nitrogen deposition are nowadays central and western Europe, eastern USA and, since the 1990s, Eastern Asia.

The deposition of atmospheric N deposition in forest stands in Europe or the USA has reached values between 20-100 kg ha^{-1} $year^{-1}$ have been reached, instead of the estimated background inputs of 1-3 kg N ha^{-1} $year^{-1}$ in the early 1900s. The term "critical load" is widely used in Europe when acceptable levels of pollutants such as sulphur or nitrogen are discussed. A quantitative estimate of the exposure to one or more pollutants below which significant harmful effects on specified elements of the environment do not occur, according to present knowledge.

The critical load of forest for nitrogen deposition has been estimated to be 10-20 kg N ha^{-1} $year^{-1}$. According to Kurz and

Rihm, estimated nitrogen deposition exceeds the critical loads in 90% of the Swiss forests. Atmospheric inputs of excess nitrogen significantly affect fine roots through effects on the plant carbon allocation pattern, biomass, secondary defence chemicals, and the storage of carbohydrates. High nitrogen contents in the soil may negatively influence the fine root mycorrhizal symbiosis and thus tree vitality might be affected.

FINE ROOTS AND NITROGEN DEPOSITION

Fine roots are important structural and functional components of forest ecosystems. The definition of fine roots is based on diameter analyses. Usually, roots <2 mm in diameter are termed fine roots. Such analyses show that the major part of the length of tree root systems consist of fine roots and that the quantity and activity of the small-diameter parts of the root systems are of prime significance regarding water and nutrient supply. Lyr and Hoffman have shown in four tree species that fine roots account for approximately 86 to 99% of the total root length, although their contribution to root weight is only marginal.

A large proportion of net primary production in sugar maple is going to roots <0.5 mm in diameter. These roots contribute a large proportion of the root biomass and are the physiologically most active part of the root system, responsible for water and nutrient uptake. Root length is a direct indicator of the potential for the nutrient and water uptake.

The contributions of fine roots to whole plant carbon budgets and to ecosystem level carbon and nitrogen cycles have gained wider appreciation. Fine root turnover has been estimated to account for as much as 33% of global annual net primary production. In trees, belowground net primary production may exceed 50% of total net primary production, with fine roots comprising a substantial part of total belowground net primary production.

Root turnover is related to precipitation and soil factors. However, the difficulties associated with studying fine roots have resulted in a paucity of data for plant roots. Nitrogen deposition above the critical load may be considered a stress factor for plants. Plants respond to natural or simulated stress factors by changes in their above-ground parts and root systems. Experimental nitrogen addition to saplings of beech and Norway spruce in young stands on acidic and calcareous soils in afforestation plots in Switzerland induced nutrient imbalances and deficiencies. These changes were significant at added nitrogen loads of 10-20 kg N ha^{-1} $year^{-1}$ after 4-6 years of nitrogen treatment.

On acidic soil, nitrogen treatment led to acute Mg deficiency whereas on calcareous soil K and P were reduced. The most frequently used methods to measure the response of root systems include measurements of the fine root biomass, the vertical distribution of fine roots, the number of root tips and mycorrhizas, the number of live and dead root tips, the chemical composition of fine roots, changes in the uptake of nutrients, etc. High nitrogen deposition is expected to produce long-lasting effects on fine-root systems and their function.

In some ecosystems, nitrogen fertilisation has negative effects on fine-root development, resulting in a 50% reduction of fine roots compared to controls. Forest trees with a reduced or damaged root system may be sensitive to environmental stress. Nadelhoffer reported that with increasing nitrogen availability, fine-root biomass decreases and turnover increases. The ratio of NH_4^+ to NO_3^+ might also influence the fine root morphology as was shown for beech by Paar.

Van Dijk et al. applied 0-48-480 kg N ha^{-1} $year^{-1}$ (as $(NH_4)_2SO_4$) to young Pinus sylvestris, Pinus nigra and Pseudotsuga menziesii in a pot experiment. After seven months the coarse root biomass had not changed, but the fine root biomass decreased by 36% at the highest nitrogen application.

In parallel, a 63% decrease in mycorrhizal infection at the highest nitrogen application was found. Matzner and Murach showed a negative relation between nitrogen deposition and soil acidification on the one hand and a reduce the fine root biomass and root length of trees.

Norway spruce trees a drastic decrease in root length per plant and also per gram root mass with high N availability.High levels of NH_4^+ deposition may have strong negative effects on root uptake capacity. In one nitrogen saturation study the fine root biomass and the number of root tips of Pinus sylvestris have increased after reduction of the present nitrogen deposition to pre-industrial levels, indicating a restricted root growth and nutrient uptake capacity at the ambient nitrogen load of ca. 40 kg N ha^{-1} $year^{-1}$.

Although a majority of studies reported negative effects, there are also reports on van Dijk et al., 1990; Clemensson-Lindell and Persson, 1995, unchanged, or even increased root production under high nitrogen deposition. This indicates that root responses to increased nitrogen deposition may depend strongly on the chemical and physical properties of the soil.

An increased biomass of fine roots in spruce after addition of ammonium sulphate at a dose corresponding to 100 kg N ha^{-1} $year^{-1}$ in the first and third year of application. In forest trees such as beech and Norway spruce, most fine roots ere associated with ectomycorrhizal fungi (EMF). Mycorrhizas play an indispensable role in the uptake of nutrients and water by trees.

This symbiosis is an important means of nitrogen uptake by many plants, since these fungi secrete extracellular enzymes that break down organic material in the soil and play a critical role in tree nutrition and carbon balance, supplying soil resources to their plant hosts in exchange for sugars. Ectomycorrhizal trees dominate nitrogen-limited forest ecosystems. Increased nitrogen inputs might be responsible for the disturbance of mycorrhiza, leading to nutrient imbalances in the plants.

Termorshuizen et al. applied 0 to 400 kg N ha^{-1} $year^{-1}$ either as ammonium or nitrate to young Pinus sylvestris, inoculated with Paxillus involutus in a pot experiment. Above application rates of 10 kg N ha^{-1} $year^{-1}$, a decrease in the amount of mycorrhizal root tips and the number of sclerotia was found.

Increased contents of nitrogen in soil induced an apparently decreased frequency of mycorrhizal roots in the humus layer from 80% to 50%. Similarly, fertilisation experiments have shown that an application of 3′100 kg N ha^{-1} suppressed the frequency of mycorrhizas from 87% to 77% and the formation of fruiting bodies by more than 50%. The number of fruit bodies and number of species of ectomycorrhizal fungi are lower in areas with high levels of nitrogenous air pollution.

Arnebrandt found that the growth of external ectomycorrhizal mycelium was negatively affected by nitrogen additions. Reduced growth of mycorrhizal mycelium resulting from high nitrogen deposition is likely to reduce the potential of the fungus to take up minerals and water and transfer them to the tree.

Boxman and Roelofs found that the uptake of cations like Mg^{2+}, K^{+}, and Ca^{2+} by mycorrhizal fungi growing in symbioses with pine seedlings decreased at high NH_4^+ levels. Branderud found already after 1.5 year a decrease in fruit body production of mycorrhizal species at a nitrogen application of 35 kg N ha^{-1} $year^{-1}$ (as NH_4NO_3) in a Picea abies stand.

POLLUTANTS AND SOIL ACIDIFICATION

Acidification caused by emissions of nitrogen and sulphur and associated adverse effects on forest ecosystems, surface waters and human health has been an issue on the political agenda for decades. The atmospheric reactions of NO_x and SO_2 lead to nitrous and sulphuric acid formation and to an increased inflow of anions and protons into soils, which can increase ion leaching from soils.

Consequently, "acidic deposition" can be defined as the inflow of hydrogen ions from the atmosphere to forests. In addition, the reactions of sulphates and nitrates in soil are considered in studies that analyse the effects of acidic deposition on forests. "Soil acidification" is a general term that is used to cover harmful chemical changes in soils due to acidic deposition, but as a phenomenon soil acidification can also occur as a result of internal proton production in soil.

According to Vanmechelen et al., 65% of european forest topsoils are acidic (pH ($CaCl_2$) = 4.5) and more than 30% of soils are desaturated (base saturation = 15%). De Vries et al. indicate that nitrogen depositions of 13 to 25 kg N ha^{-1} $year^{-1}$ lead to elevated NO_{3-} leaching, decreasing base cation pools in soil, and increasing soil acidity. Many forest soils have become acidified and have high levels of both Al and NO_3 in the soil solution.

Soil acidification has accelerated also in Switzerland. The concern on acidification effects has lead to the definition of critical loads for acid deposition which take into account the buffering capacity of the soil and the ecological effects.

Soil Acidification

Soils acidify naturally as they weather over thousands of years. The acidity of any soil varies according to the type of rock it comes from, the length of time it has weathered, and the local climate. Hence, some soils are naturally acidic while others are more alkaline. Agricultural practices can greatly increase the rate of acidification. Changes in soils that may occur in acidified soils include: decreased base saturation, increased availability of aluminium andmanganese, and competition between base cations and Al for exchange sites in the fine roots.

Majdi and Persson reported that the fine roots are considerably influenced by the degree of acidification. The effects of soil acidity depend on the investigated parameter. Total root biomass may be increased in acidic soils. Godbold et al., 2003 investigated the fine root distribution and turnover in ca. 40-year-old Norway spruce stands, growing on four sites that differed in soil acidity. At one of the most acidic sites, fine root density in the humus layer was more than twice as high as at the least acidic site.

Most of the roots and the root tips were found in the humus layer and in the first mineral soil horizon (0- 10 cm). There was a significantly different decrease in specific root length (cm g_{DM}^{-1}) and specific root tip density (root tips g_{DM}^{-1}) in the more acidified stands. However, Leuschner et al. reported that beech fine root tip density (tips per soil volume), were much higher in acidic than in basic soils. Recently, it has been shown that subsoil acidity affects the distribution of fine roots of Norway spruce, confirming ideas expressed in earlier work.

The uptake and transport of several essential elements i.e. Ca, Mg, K and P may be adversely affected by Al. In acidic soils Al is soluble and acts as a phytotoxic agent and may be an important growth factor for forest trees. Inhibition of root growth and changes in root morphology are the principal visible symptoms of the Al-toxicity. It has been hypothesized that increased Al solubility at low pH could damage the roots of the trees and impair nutrient and water uptake.

SECONDARY PLANT METABOLITES

Plants are known to produce a large number of low molecular weight secondary compounds. Due to the remarkable development of analytical methodology; e.g., in gas chromatography (GC) and high-pressure liquid chromatography (HPLC) coupled to mass spectroscopy (MS), it has been possible to characterise even minor components in plants.

Previously, most of these compounds were considered metabolic waste they did not seem to have any clear function in the organisms that produced them. However, more recently, such compounds have been shown to exhibit diverse biological functions. Many secondary compounds are now known to be of ecological importance, not only due to their role as protectants from effects of UV-light and other abiotic factors, but also as a defence against herbivores and pathogens.

In contrast to primary metabolites, secondary compounds vary widely in their distribution among plant species. Some of these compounds occur only sporadically, whereas others are distributed widely throughout the plant kingdom. In addition, the intraspecific variation in plant secondary metabolite composition is often considerable.

Such variation may also occur at the level of an individual plant; i.e., secondary products are not found uniformly throughout the plant but are often limited to particular organs, and even to particular cells and tissues within that organ. Two widely distributed groups of secondary compounds that are present in all plants are phenolics and terpenoids. In the following, only the phenolic compounds will be addressed.

Phenolic Compounds

A phenol is a chemical compound with at least one aromatic ring bearing one or more hydroxyl groups. Generally, these compounds occur as different derivatives formed by condensation or addition reactions, thus contributing to the wide variety of chemical compounds found in plants. Phenolic compounds have many functions in plants. They are important in the defence mechanisms of plants under different environmental stress conditions.

The chemical defence strategy used by trees against atmospheric stress and herbivory includes the synthesis and accumulation of secondary metabolites, such as phenolic compounds. Phenolics can provide resistance for plants against fungal, bacterial, and viral infections. The compounds of interest in the present thesis were 4-hydroxyacetophenone, piceatannol

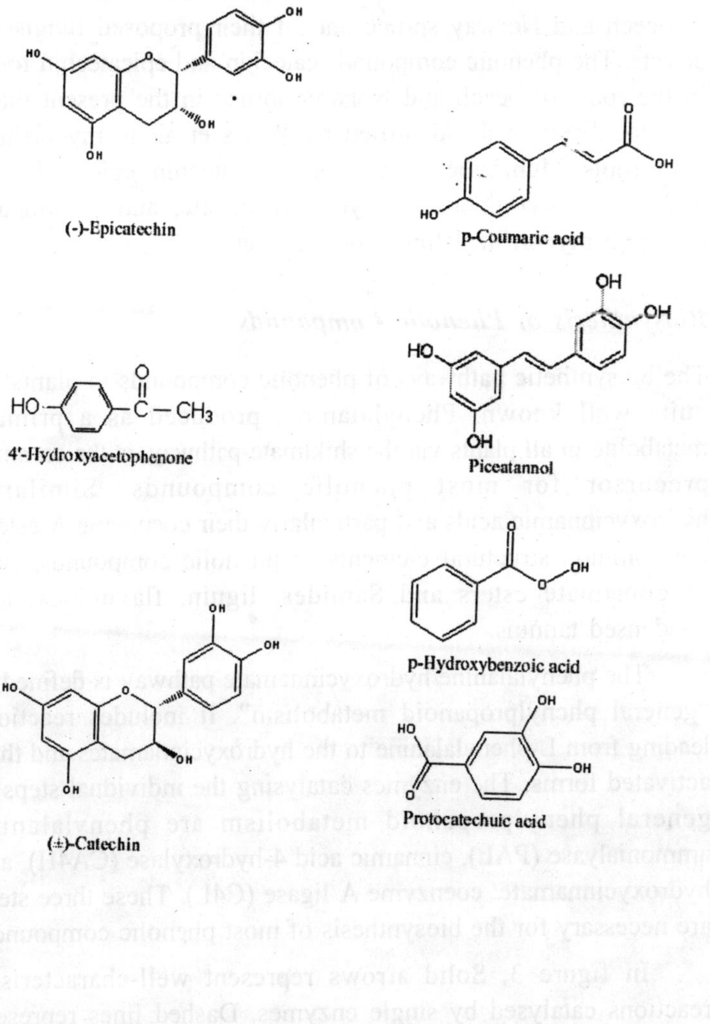

Figure 2: Structure of phenolic compounds.

(3,3', 4,5'-tetrahydroxy-transstilbene), p-hydroxybenzoic acid, protocatechuic acid (3,4-dihydroxybenzoic acid), p-coumaric acid (4-hydroxycinnamic acid), (-)-epicatechin and (+)-catechin in fine roots (Figure 1).

These compounds were selected because of they occurred in beech and Norway spruce and of their proposed fungistatic effects. The phenolic compounds catechin and epicatechin found in the roots of beech and Norway spruce in the present thesis had been previously identified by Weiss et al. in mycorrhizal larch roots. Munzenberger et al. found catechin, epicatechin, 4-hydroxybenzoyl-glucose, 4-hydroxybenzoate, and additionally picein in mycorrhizal fine roots of larch.

Biosynthesis of Phenolic Compounds

The biosynthetic pathways of phenolic compounds in plants are quite well known. Phenylalanine, produced as a primary metabolite in all plants via the shikimate pathway, is the common precursor for most phenolic compounds. Similarly, hydroxycinnamic acids and particularly their coenzyme A esters, are common structural elements of phenolic compounds, such as cinnamate esters and Samides, lignin, flavonoids, and condensed tannins.

The phenylalanine/hydroxycinnamate pathway is defined as "general phenylpropanoid metabolisın". It includes reactions leading from L-phenylalanine to the hydroxycinnamates and their activated forms. The enzymes catalysing the individual steps in general phenylpropanoid metabolism are phenylalanine ammonialyase (PAL), cinnamic acid 4-hydroxylase (CA4H), and hydroxycinnamate: coenzyme A ligase (C4L). These three steps are necessary for the biosynthesis of most phenolic compounds.

In figure 3, Solid arrows represent well-characterised reactions catalysed by single enzymes. Dashed lines represent transformations that require multiple enzymes that are less characterised, or vary among plant species. Enzymes: CA4H, CHS, 4CL, PAL.

Beech leaves phenolic compounds be decreased by nitrogen fertilisation. Probably because of a higher demand for carbohydrates during absorption and assimilation of nitrogen, the phenolic concentration of birch leaves decreases. The composition and concentrations of phenolic compounds were studied from Keski-Saari and Julkunen-Tiitto in the first true leaves, cotyledons, stems and roots of seedlings of mountain birch.

The differences in secondary compounds among these plant parts were both qualitative and quantitative. The seedlings were growth at three levels of nitrogen supply and the effect of nitrogen on concentrations of phenolic compounds was studied. The concentrations in roots were highest at low-N. In nitrogen fertilisation experiments, carbon allocation between different metabolic pathways and into individual phenolic compounds differs considerably from each other.

Environmental conditions, such as the fertility of the habitat, also affect the concentrations of secondary compounds. Nitrogen availability is thought to affect the concentrations of secondary compounds in a predictable manner: The plants growing in nitrogen-poor conditions are thought to have more secondary compounds than plants growing in nitrogen-rich conditions. The few studies available comparing phenolic compound production in plants growing on a acidity gradient in natural conditions have found higher phenolic concentrations in plants associated with lower pH.

This correlation is associated with lower N availability found in acidic soils although some studies have found direct effects of pH. According to the carbon/nutrient balance hypothesis, nitrogen availability in the soil is low, the low resource availability limits the growth of plants more than the photosynthesis, and plants allocate their extra carbon that they cannot use for growth to carbon-based secondary compounds.

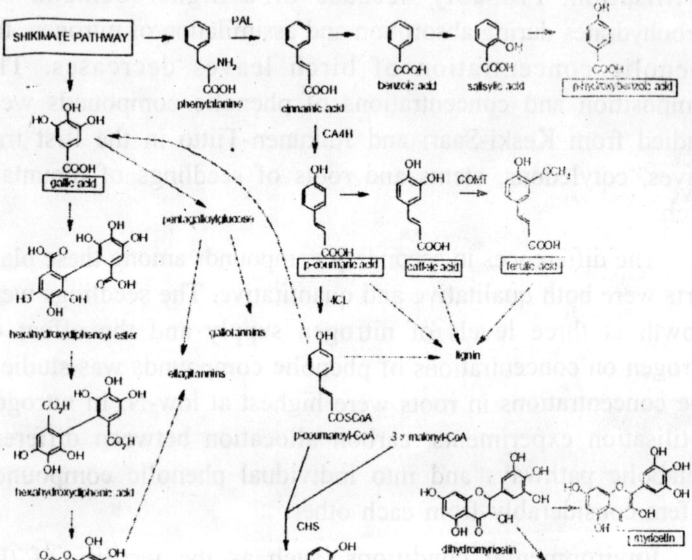

Figure 3: Biosynthesis of hydroxycinnamic acids, hydroxybenzoic acids and flavonoids.

Furthermore, the predictions of the theory seem more reliable in some plants species a other, and the carbon/nutrient balance hypothesis can perhaps only be usefully applied to specific groups of secondary compounds and particular plant species. According to the growth/differentiation balance hypothesis, nitrogen or carbon availability are not the only environmental conditions that influence the secondary concentrations in a predictable way, but the trade-off happens between all growth and differentiation processes.

The protein competition model predicts that protein synthesis and phenolic synthesis are alternative competitive demands and, for example during the development of leaves the demands for growth and protection alternate. A possible regulatory step between growth and the synthesis of secondary metabolites are the phenolic amino acid, phenylalanine, and is used for synthesis of both proteins and of secondary compounds produced by the shikimic acid pathway.

The hypotheses predict increases in the concentration of carbon-rich compounds when nutrients are less available. A reduction in the levels of carbon/based secondary compounds in response to fertilisers have been observed experimentally for many plants. Shaw et al. reported unaffected total phenolic and tannin levels in the root collar area of Douglas-fir seedlings by high nitrogen treatments.

Pasqualini et al. showed a significantly negative correlation between total phenolics in needles of Aleppo pine from 6 sites affected by various forms of atmospheric pollutants and nitrogen oxide concentrations. This negative correlation may be explained by the positive impact of these pollutants on the activity of nitrate reductase. This enzyme promotes nitrogen assimilation and several studies have shown the same negative correlation in needles or leaves of various species.

The correlations between air pollutants and simple phenolic compounds showed concordant results. Several enzymes, which take part in the shikimic acid biosynthetic pathways, may be more sensitive than others in relation to different pollutants. Increased infection of Douglas-fir (Pseudotsuga menziensii (Mirb.) Franco) by Armillaria ostoyae Herink occurred after fertilisation with nitrogen.

Conifer species with increased susceptibility to Armillaria ostoyae generally have fewer phenolic compounds and more sugar in root bark, suggesting that the ratio of available sugars to phenolic compounds may be related to the success of the fungal infection.

Phenolic Compounds and Plant Pathogens

Plants are capable, through their secondary metabolic products to reduce the attack of pathogenic organisms and phenolic compounds may be an important group of secondary metabolites involved in resistance to pathogens. It has long been known that fungal secondary metabolites are crucial to the pathogenicity of many fungi. Plant pathologists have traditionally recognized two types of antiinfectional defence chemicals in plants: those that are present constitutively and those where de novo synthesis is induced as a result of an infection.

These two groups are thus pre- and post-infectional defences, respectively. The content in the needles of Picea abies of 4-hydroxyacetophenone and its glucoside Picein have been proposed as stress indicators. Recently the phenols get into the centre of attention as the potential defensive compounds of plants. The proanthocyanidins have been shown to function as preformed defence metabolites in their hosts.

When the concentration of proanthocyanidins decrease, Botrytis cinerea becomes pathogenic. In the case of pathogenic plant-fungal interaction, the fungus Heterobasidion annosum catabolizes catechin and causes a reduction in this compound in infected Pinus roots. The fungistatic properties of phenolic compounds have been documented in vitro. Terpenes and phenolics showed an inhibitory effect to pathogens at concentrations observed in pines and other conifer species.

Blodgett and Stanosz also examined the effect of phenolic compounds and monoterpenes on mycelial growth of Sphaeropsis sapinea in vitro. The bioassays revealed that pinosylvin and several monoterpenes occurring in red pine inhibit mycelial growth of this pathogen. The concentrations of phenolics in the bark of American beech trees resistant to infection by Nectria cocconea var. faginata, remained high even 6 months after the attack.

The effect of an extract of Norway spruce bark on the growth of the root rot fungus Heterobasidion annosum and the blue-stain fungus Ceratocystis polonica was investigated in vitro by Evensen et al. Heterobasidion annosum was not negatively affected, the extract had only fungistatic effects on the blue-stain fungus. The production and accumulation of phenolic compounds are often associated with injury responses and the wound healing of plants, but these substances also have fungitoxic activity.

Previous studies of host-fungus interactions in the xylem of living trees have strongly emphasised the importance of host responses to fungal invasion. By contrast, the mechanisms underlying reaction zone penetration and subsequent lesion expansion by wood decay fungi are poorly understood. Recently, however, Schwarze and Baum described in detail a variety of mechanisms by which reaction zones of beech wood were penetrated in vivo or when challenged in vitro with decay fungi. When attacked by bark beetles and their associated fungi, conifers form a necrotic reaction zone around the site of infection.

This reaction zone becomes impregnated with secondary metabolites such as phenols and terpenes. Métraux related phenolic accumulation in plants to defence mechanisms against pathogens. After wounding or infection of Norway spruce, there are both quantitative and qualitative changes in the phenol content in the reaction zone. These changes, together with the presence of polyphenolic parenchyma cells in the phloem of Norway spruce, suggest an important role of phenols in coniferous resistance reactions.

The polyphenolic parenchyma cells are located in continuous sheets in the secondary phloem and contain dense vacuolar deposits of phenols. In vitro bioassays have demonstrated a fungistatic or fungitoxic effect of some of the phenols normally found in the sapwood and phloem of Norway spruce. In 1940, K.O. Müller and H. Börger proposed the phytoalexin theory of disease resistance in plants which offered an explanation of certain aspects of disease resistance based on the postinfectional formation of warding-off-compounds by the plant.

One of the best and extensively-studied defence responses of plants to pathogen infection is the induced accumulation of secondary metabolites such as phytoalexins. Phytoalexins are a diverse group of low molecular weight anti-microbial compounds that are synthesized and accumulated in appreciable amounts in plants after stimulation by the various types of pathogens or by chemical or mechanical injury and are toxic to pathogens.

Nitrogen Fertilisation and Pathogens

Nitrogen availability can affect a plant's susceptibility to pathogens. Van Dijk et al. found a relation between the availability of nitrogen and the incidence of damage caused by Sphaeropsis sapinea (Fr.) Dyko & Sutton. Trees affecte more strongly by Sphaeropsis sapinea contained more nitrogen. The present investigation was made in a nitrogen addition experiment designed to evaluate critical loads for nitrogen.

Fertilisation with 0-160 kg N ha^{-1} $year^{1}$ caused significant changes in nutrient concentrations and ratios in beech in afforestation plots in Switzerland and trees were significantly more infested by Apiognomonia errabunda, Phomopsis sp., Phyllaphis fagi and Botrytis cinerea when fertilised with nitrogen. However there are scarce informations about the effect of nitrogen on the interaction between roots and root pathogens.

Two-year old plants of Pinus nigra were grown for three years in pots and fertilised with 5 treatments of ammonium sulphate, in combination with two levels of potassium sulphate. The 5-year-old plants were then inoculated with Sphaeropsis. The bark necroses were much higher in the plants treated with ammonium sulphate, whereas in the plots with high levels of potassium sulphate the necrosis was low.

Effects of ammonium sulphate upon fungal damage were already observed at an addition of 75 kg N ha^{-1} $year^{-1}$, but were very significant in the plants treated with 150 kg N ha^{-1} $year^{-1}$. The influence of nitrogen on the disease seems to depend on the host-pathogen combination and the form of soil nitrogen. A stimulation of in vivo production of sporangia by several Phytophthora species, with increasing concentrations of nitrate in the soil leachate. This stimulation was probably a major cause of the increasing difference in the fine-root length and the number of fine root tips between uninfected and infected Quercus robur seedling with increasing nitrate concentration. In field studies, pH as well as nitrate and Ca concentrations were significantly higher in soils infested with Phytophthora than in uninfested soils.

SITE-SPECIFIC NITROGEN FERTILISATION

In crop production, nitrogen and water are usually the major growth limiting factors. Of all nutrient amendments to Swedish agricultural soils, N fertilisation has by far the greatest effects in terms of increasing crop production. Nitrogen fertilisation affects not only yield, but also the quality of cereals and other crops in several ways. The amount and timing of plant available N is crucial for the grain crude protein content (CP) in cereals.

The desired level depends on the final use of the product. For example, high CP is desired in bread wheat, but low CP in biscuit wheat and intermediate CP concentrations in malt barley. Too much N may lead to lodging, which in turn reduces grain quality due to over-moist conditions, with germination in the ear as a consequence. Therefore it is important for the economics of crop production to apply the economically optimum N rate.

Fossil Resources Versus Recycling: The production of mineral nitrogen fertilisers requires large amounts of energy, and therefore involves consumption of fossil fuels, which also adds to the 8 greenhouse effect. However, the use of mineral fertilisers should not only be limited due to their energy requirements. They should be regarded as a complement to the N resources that already exist, i.e. plant available N from the soil, preceding crops and organic fertilisers, such as manure and other waste products.

Organic fertilisers, including organic wastes from society, constitute a valuable source of nutrients for fertilisation, but may cause eutrophication problems if these nutrients leach to the environment instead of being utilized in crop production. If all nitrogen that leaves the fields could be recycled and applied to the fields in such a way that it became available to the crop during growth, less mineral N fertiliser produced by industry would be necessary. However, some N will be lost thorough leaching in sewage treatment plants or through gaseous emissions from manure and other organic wastes before they are applied to the soil and some N will be mineralized after the crop growing period and then be subject to leaching and denitrification.

Human wastes are generaliy not applied to agricultural fields at all, due to problems with, or fear of, contamination by hazardous chemicals and pathogens. As long as these problems remain, we need mineral N fertiliser to complement organic fertilisers, but methods to economize on all N sources are essential.

Nitrogen as a Mobile Element: Nitrogen circulates between the atmosphere, the soil and living organisms, and many factors and processes are involved in this N turnover. Only a very small proportion of the N in soil is directly available to plants, mainly in the form of NO_3^- and NH_4^+ ions. As nitrogen is a very mobile element, adequate amounts of fertiliser must be applied in each growing season, since excess amounts will not remain available to any large extent for the following growth period. Nitrogen must also be supplied at appropriate development stages of the crop to achieve the desired effects on yield and quality. Early N application primarily promotes vegetative growth and grain yield, whereas N available later in the growing season mainly favours grain filling and protein production.

Weather Dependency: Yield response to fertiliser N varies depending on plant available soil N (Np) and potential yield, which in turn depend on a number of factors. Np is here defined as the amount of soil N that the crop takes up during the growing season. Both yield formation processes and Np are very weather dependent. As weather varies between years and is difficult to predict, the task of predicting N fertiliser demand in advance is very difficult.

Spatial Variability: Since fertiliser N demand, as influenced by yield level and by quality requirements as described above, is dependent on many factors that can vary within a field, the N fertilisation demand may vary spatially. When several influencing factors vary spatially, the variation in N demand can be difficult to predict.

SITE-SPECIFIC N FERTILISATION

To achieve the maximum net returns of added N, adequate quality and minimized losses to the environment, site-specific adjustment is necessary if the within-field variations are considerable. The development of such techniques, together with for instance better utilization of manures, efficient recycling of nutrients from society, the use of catch crops, etc., can be regarded as a measure for reducing nitrogen losses and thus protecting the environment. For this, models helping us to decide when and how to consider spatial variability are needed. In principle, traditional methods should still be valid when applied to smaller surface units. However, as these methods may involve laborious and time-consuming collection of data that is unrealistic to perform densely within the field, new approaches may be necessary.

The aims of the experiments were to quantify spatial variation in crop response to fertiliser nitrogen and to identify relationships between soil characteristics, plant available soil N and potential grain yield. Since water is the most common factor limiting yield apart from nitrogen in most agricultural areas, relationships between yield and plant available water and related soil characteristics within the field were investigated. The objective was to find out whether such relationships could be used in models for the prediction of fertiliser N demand based on within-field variations in plant available soil water and nitrogen. N fertilisation demand is dependent on potential grain yield and plant available soil N. If these parameters are significantly related to soil characteristics within a field, the relationships could be used for estimating fertilisation demand site-specifically.

Site-specific Crop Management

Site-specific crop management or Precision agriculture aims at matching inputs and agronomic practices with soil attributes and crop requirements as they vary across individual fields. Such differential treatments within fields are in contrast to the uniform treatments that characterize current management systems. The size of agricultural fields has increased during the past century through fusion of smaller plots and removal of open ditches and other obstacles. With the larger agricultural units of today, it is more difficult for farmers to maintain knowledge of specific details concerning their fields. At the same time, the development of information technology opens up new possibilities to handle large amounts of information.

Precision agriculture is a way to again divide individual fields into smaller units and better adjust the management to the conditions of each unit. The aims are to optimize yield and quality of the crop, save inputs and improve profits. More economical use of inputs, such as fertilisers and pesticides, may not only be profitable for the farmer, but may also reduce undesired losses of substances to the environment.

Precision agriculture was initiated in the mid-1980s, In the USA, using newly available technology to improve the application of fertilisers by varying rates as needed within fields. Since then, the concept has been adapted to a variety of practices and crops in many countries. However, several challenges limit a broader adoption, including economic, agronomic and technological barriers. For instance, high costs are incurred in the collection of required data and in the acquisition of technical devices and there is some uncertainty regarding whether precision agriculture leads to increased returns.

The potential benefits of applying N fertiliser at the optimum rate for each soil type in a spatially variable field can be positive but small and with a great variability between seasons. Another reason for not adopting precision agriculture is the lack of qualified agronomic services for support. Advisors must possess tools and knowledge to interpret yield and soil maps and to produce maps for variable rate application. As the supply of qualified agronomic services increases and costs decrease, e.g. by better sampling schemes and cheaper technical devices, implementation may increase.

Within-field Variations in Yield: Recording differences in grain yield within individual agricultural fields using yield monitors on the harvester was facilitated when the global positioning system (GPS) became available. The size of the yield variation varies between fields from small, more or less random, variation to up to several thousands of kg grain ha-1, which in many cases is related to the range of soil texture variation. Yield is largely dependent on soil moisture conditions, which in turn are correlated to soil texture.

The magnitude of the variation in yield may also vary between years. In fields where yield variation is highly influenced by differences in the incidence of drought, yield variation is smaller in moister years and larger in dry years. Topography also affects yield by its effect on soil moisture and temperature, which for example influence crop development, e.g. germination, tiller production and grain filling. Yield losses may occur on north-facing land whereas south-facing slopes increase biomass production. Yields are also influenced by weeds, pests and diseases.

In the early days of yield mapping, there was much hope that such maps could be used as a direct measure for e.g. fertiliser requirements within the field. However, yield maps vary between years and should be interpreted to identify causes of variation before they are used for adjusting management practices. By identifying low yielding areas and the reasons for limited yields, different types of measures can be taken. Inputs can be saved in low yielding areas by decreasing the fertiliser rates. On the other hand, measures such as drainage or liming may be needed in those areas to increase the yield.

Soil nutrient conditions can also vary within the field. Yield maps together with soil nutrient maps indicate where more or less fertiliser is needed. A high yield potential and low soil nutrient content indicate the need for larger amounts of fertiliser. Several models have been proposed based on either yield maps, soil maps or both, or on multivariate methods including several yield limiting factors instead of yield itself. Another approach includes models based on the current nutrient status of the crop, to be used when fertilisation is performed in a growing crop.

The latter is especially interesting for N application, since the seasonal variation in N demand can be considerable.

For P and K, fertiliser requirements are not as sensitive to seasonal variations. Variable P application that considers yield responses and soil P will even out differences in soil P slowly, unless much larger P additions than required by the crop are applied on areas with low P contents. Some researchers simulated a 50% decrease in variability over 20 years. Another some researchers could not detect any significant grain yield increase within the first years when they compared variable and uniform application rates of lime, phosphorus and potassium. Several experiments have tested P and K fertilisation based on topography, which seems to be profitable in areas where yield potential is dependent on elevation.

Variable rate application of lime can be based on pH maps complemented by clay and soil organic matter (SOM) maps. The contents of clay and SOM, which can be considered to influence the cation exchange capacity of a soil, thus affect the amount of lime required to reach a certain target pH.

Site-specific Nitrogen Fertilisation Demand Estimation Methods

Soil N Analyzes

Plant available soil N during the growing season can be divided into two components: soil mineral N (NH_4^+ + NO_3^-) in the root zone in early spring and N mineralized during the following growing season. In several countries the overwintering store of mineral N may be considerable and constitute a basis for N fertilisation recommendations. However, this involves soil sampling, which is costly and time-consuming and is therefore not feasible for site-specific farming where information at many locations within the field is required.

Several rapid techniques for the determination of soil mineral N exist. One of them, a nitrate selective electrode, has a fast and accurate on-the-go field unit, but requires tedious calibration. Another method is based on near infrared reflectance (NIR) of the soil, which could predict nitrate levels after field calibration. However, nitrate levels in spring in northern Europe are often very small in relation to the amounts of N mineralized later during the growing season. In addition, NIR is not directly affected by nitrate, but only by other related constituents. More interesting is perhaps that NIR in a similar way can be used to estimate plant available soil N during the growing season (Np) in fields with large variation in SOM content. NIR can be a better predictor of Np than SOM content (Figure 4), probably because NIR may be affected not only by SOM content, but also by the quality of SOM, as well as by other soil properties that affect N mineralization.

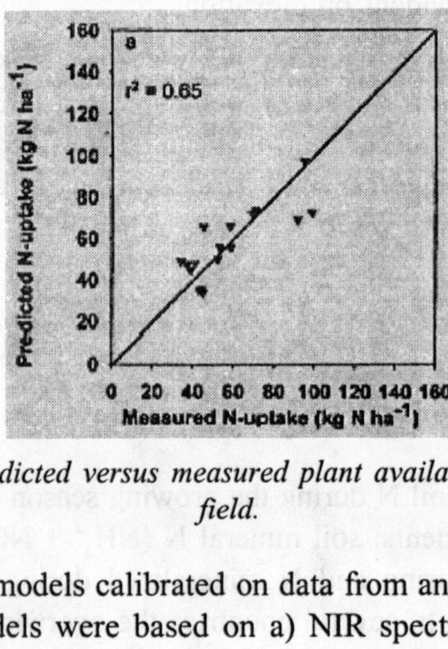

Figure 4: Predicted versus measured plant available soil N in a field.

In this figure, models calibrated on data from another field were used. The models were based on a) NIR spectra and b) SOM data.

Remote sensing for estimations of plant N status

Remote sensing is the art of obtaining information about an object through the analysis of data acquired by a device that is not in contact with the object under investigation. Remote sensing, e.g. image and spectral analyzes, is used in precision agriculture to map differences within fields. Image analysis makes it possible to map weeds and spectral analysis can be used to map fungal infections, plant density, nutrient status of the crop and, as mentioned earlier, for the description of soil properties. Plant density and nutrient status of crops are of significant importance for fertiliser demand and therefore spectral analysis can be used to estimate fertiliser N demand within the growing season.

The proportions of radiation reflected, absorbed and transmitted at different wavelengths vary for different features, depending on material and condition. When a plant 13 canopy grows, the bare soil reflectance spectrum is progressively replaced by the reflectance spectrum of the plants. Thus, the higher the leaf density during growth, the lower the visible reflectance and the higher the near-infrared reflectance (NIR) until a saturation level is reached when the soil is completely covered by the crop. In addition, N deficiency changes the reflectance pattern (Figure 5).

As a result of insufficient N supply, the visible reflectance is increased due to decreasing chlorophyll content and the NIR is decreased due to decreasing number of cell layers. Reflectance of selected wavelength bands within the visible and the near infrared spectrum is used in different indices to estimate biomass and N concentration and thereby fertilisation demands.

Variable rate applications of N based on real-time assessments of canopy size have been successful, resulting in more efficient use of N, and thereby better N economy and reduced N surplus. The principle of using biomass to estimate crop N demand can be used in several stages of plant development at which fertilisation is performed. For fertilisation at tillering, the smaller the plant density, the more N is needed to achieve optimal shoot density for larger yield. At a later stage, however, a small biomass indicates low potential yield, and fertilisation should therefore in principle be decreased. However, decreased biomass is often associated with N deficiency and in that case fertilisation should be increased, as it is in the concept of the Yara N-sensor.

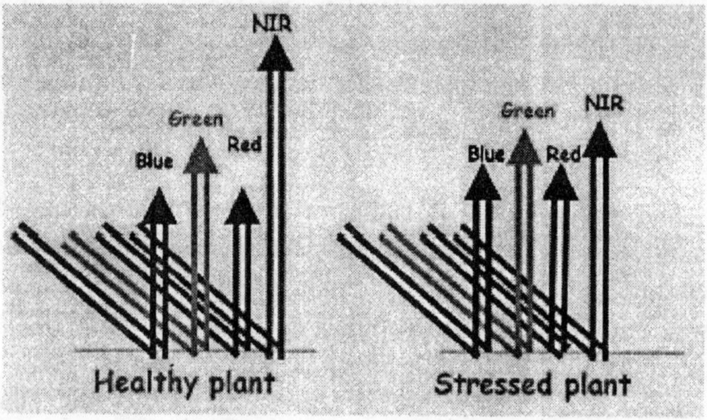

Figure 5: Schematic illustration of reflected light within different wavelengths for a healthy plant and a stressed plant.

With very large biomass there is a risk for lodging if too much N is applied, which motivates decreased N fertilisation. At very small biomass, N fertilisation is not considered to be able to compensate for N deficiency and the N fertilisation at biomass lower than a certain value is therefore not increased any further. Reflectance data can be collected either with aerial photographs or with sensors mounted on a tractor. Sensors on the tractor provide the opportunity for on-line application directly during measuring, and the technique is less dependent on suitable weather conditions than aerial photography.

Characteristics of mapping soil

Soil characteristics, such as pH and contents of nutrients, clay and soil organic matter, may traditionally be mapped through soil sampling at various points in the field. To achieve data coverage of the whole field, the measured data must be interpolated. However, soil sampling is time-consuming and expensive and therefore available samples are often few and sparse, which in turn makes the interpolated data unreliable, especially in small fields where too few samples are taken to check if there is any spatial correlation between observations. There must be a spatial correlation between neighbouring observations to motivate an interpolation. Spatial correlation is described in variograms created from raw data prior to interpolation.

The variogram is necessary for geostatistical interpolation methods such as kriging, whereas other purely mathematical methods such as inverse distance weighting are possible to carry out without any variogram, even though it gives good hints about what parameter settings to use. The variogram (Figure 6) is characterized by its nugget, sill and range.Often a large number of observations are needed for reliable estimates of variograms used for checking geostatistical relations prior to interpolation.

50-100 samples is far from the number of soil samples that it is feasible to collect within a normal-sized field in most countries. Other methods for quick and cheap collection of dense data have therefore been tested to replace or complement traditional sampling methods. Variables that can be measured at great density are for instance yield, soil electrical conductivity (SEC), topography and crop or soil reflectance.

These can be used for targeting sample points, improving interpolation or in some cases as a direct measure if correlated with the variable of interest.

Interpolation can be improved by cokriging, which is a method of interpolating sparse spatial information in cases where correlated parameters are measured at greater density. Targeted sampling represents a way of reducing sample size and hence costs. Various approaches have been proposed. Sample points are selected within grid cells at locations where a SEC map created from few sample points deviates most from a SEC map created from all sample points. Also use an algorithm to select appropriate calibration sites from dense SEC data, to be used for regression-based prediction of soil salinity. A geostatistical approach involves the variance quad tree, where uniform areas are sampled more sparsely and areas with large variation more intensively.

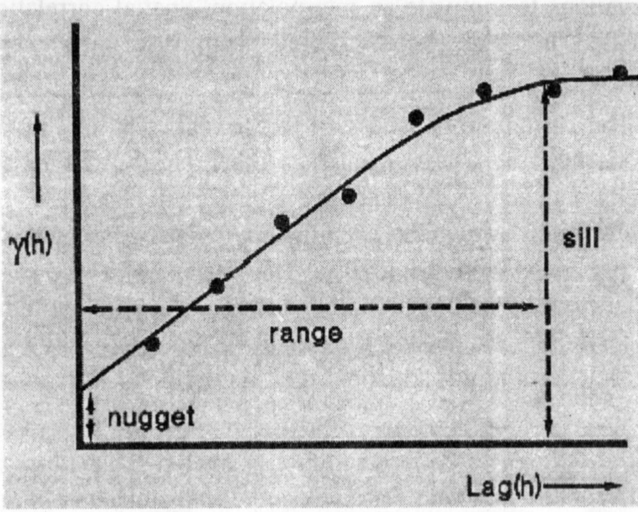

Figure 6: Schematic example of a variogram with range, nugget and sill, where γ(h) is semivariance and Lag(h) is the distance between sampling points.

In image classification, satellite images are used to divide the field into zones within which several subsamples are bulked together to get one sample representing the whole zone. Such zones can be managed homogeneously, and are often called management zones. These may also be created, for instance, on the basis of yield data, SEC or both . Varying management between a few management zones instead of adjusting it to a large number of small grids not only requires fewer soil samples during data collection, but also less advanced technology for management. It should be considered whether to use grid information, management zones or a field average for each field individually, based on the degree of variation and the potential to discern a few homogeneous areas.

OPTIMUM NITROGEN FERTILISATION PRINCIPLES

Yield responds positively to N fertilisation up to a certain limit, at which some other factor, e.g. water, begins to limit crop growth. Too much N may reduce yield due to lodging or more severe fungal infections. This may result in a negative response at higher N levels. The effect of N on yield can be described by response curves expressed as mathematical functions (Figure 7). According to some researchers that, the response is best represented by curves with a sharply rising portion, a turning portion and a portion where yield changes little or slowly decreases. Data presented here were found to be best described by second and third degree polynomials, square root polynomials or the Mitscherlich exponential equation.

NET NITROGEN MINERALIZATION PRINCIPLES

Fertiliser demand is dependent on the amounts of mineral N supplied to the crop from the soil. The more N the soil supplies to the crop, the less fertiliser N is needed to reach a certain yield level. As an example, economical rates of fertiliser N were calculated on the basis of 39 experiments in Sweden with spring barley, to be reduced or increased by 0.7 kg ha^{-1} as soil mineral N in spring within 0-90 cm depth increased or decreased by 1 kg ha-1.

In addition to mineral N present in the soil at the time of fertilisation in spring, N mineralized during the following growing season also contributes to the N supply of the crop. Net N mineralization is usually defined as N mineralization minus N immobilization. N mineralization refers to the transformation of organic N to mineral N (Figure 8).

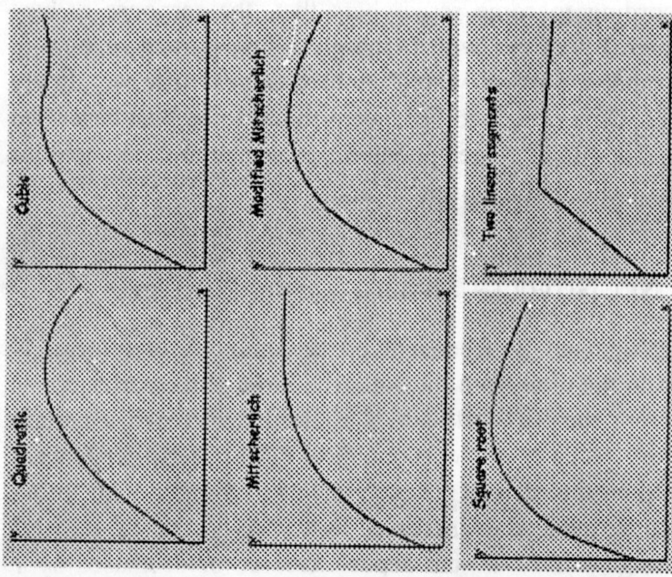

Figure 7 : Different functions describing yield response to fertilizer N.

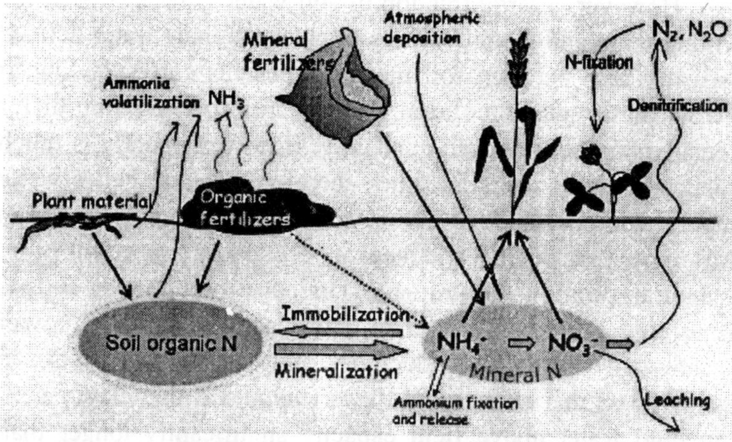

Figure 8: The N-cycle in the soil-plant-atmosphere system

The process is performed by soil microorganisms that utilize nitrogenous organic substances as an energy source. N immobilization is defined as the transformation of inorganic compounds into the organic state. Soil organisms assimilate inorganic N compounds and transform them into organic N constituents of their biomass. However, there are other processes that also affect the size of the soil contribution of N to the plant. The inorganic N pool is increased through N fixation and atmospheric deposition and decreased through denitrification, ammonium fixation, leaching and ammonia volatilization (Figure 8). Net N mineralization (Nm) is defined as soil N taken up by plants (Np) plus residual soil mineral N at harvest minus soil mineral N in spring.

PLANT WATER SUPPLY PRINCIPLES

When yield response to increasing amounts of N ceases, some other factor becomes yield limiting. This factor is frequently water. Unless irrigation is used, potential yield is reached and there is no point in adding more N. The potential water supply from soil to crops is determined by soil texture, rooting depth and groundwater level. Soil texture affects plant available water by its influence on the water holding capacity of the soil. The higher the clay content, the larger the water holding capacity.

However, a large fraction of soil water is held beyond the wilting point and the differences in plant available water defined as water held between soil moisture tensions corresponding to 1-150 metre water column (mwc) therefore do not differ considerably between sand and clay. However, rooting is much deeper in clayey soils and they therefore provide better water supply than coarser soils in dry weather conditions. Changes in soil moisture during the season are due to precipitation, evapotranspiration and drainage. Drainage through clay soils is generally much slower than from sand, due to a lower unsaturated hydraulic conductivity, unless there are cracks in the clay soil that allow rapid flow. This means that clayey soils can hold water above field capacity significantly longer than sandy soils.

OPTIMUM NITROGEN FERTILISATION ESTIMATION

To create yield response curves, fertilisation trials with several application rates and replicates are required. On this basis, economically optimum N fertilisation can be calculated by identifying where the slope of the curve equals the ratio between fertiliser and grain prices (Figure 9). Figure 6 shows a yield response curve fitted to experimental data on yield at different fertilisation rates, where optimum N fertilisation rate and corresponding yield can be estimated from the point where the slope of the curve equals the ratio between grain and fertiliser prices. Fertilisation above the optimum implies larger expenses for the extra fertiliser than the income it generates.

Similarly, fertilisation below the optimum also leads to missed returns from the yield increase that somewhat more fertiliser would have generated. Each plot in an experiment needs a net space of at least 10 m in length if harvested with a plot combine according to some recommendations for field experiments in Sweden, thus corresponding to about 25 m^2. This leads to large areas for trials with several treatments and replicates, which means that measuring within-field variations in optimum N fertilisation with this method has severe limitations.

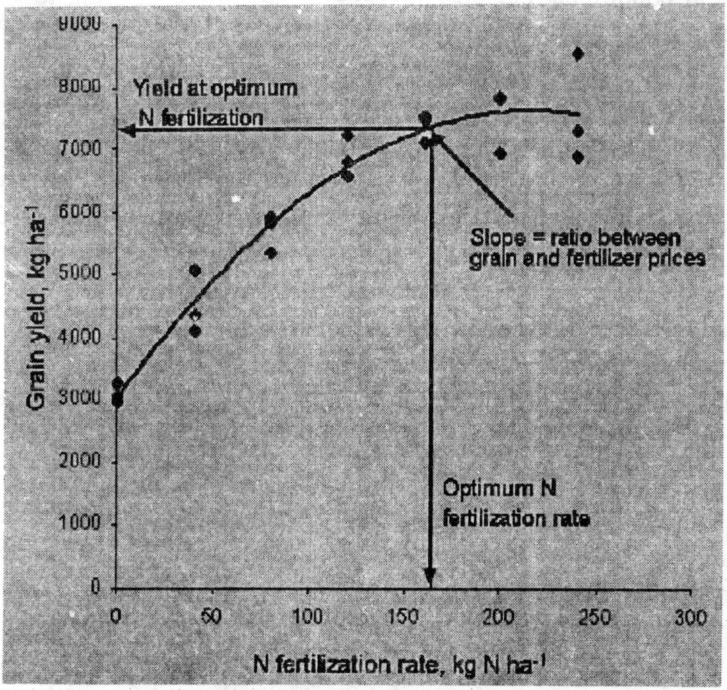

Figure 9: A yield response curve

There may be equally large variations within and between the experimental areas, and consequently much of the variation found between sites will be considered statistically insignificant, since variations within the areas will be regarded as experimental errors.

NET NITROGEN MINERALIZATION ESTIMATION

Net N mineralization (Nm) can be studied by determining mineral N accumulation in a soil incubated in controlled climatic chambers. This may give information about the potential mineralization of a soil, which may be of interest when comparing, for instance, the fertility of different soils. With the goal of developing rapid methods for the estimation of Nm under field conditions, not only incubation methods but also a large number of N availability indices based on chemical extraction methods have been developed.

However, for measuring the quantity of N mineralized in a soil during a certain period under field conditions, laboratory measurements are unreliable, since mineralization is largely affected by conditions in the field, which cannot be replicated in the laboratory. Moreover, soil preparation before incubation or chemical extraction affects release conditions. One way to study N mineralization is to measure the change in soil mineral N in the rooting zone in uncropped plots covered with a 'roof' that prevents precipitation infiltrating through the soil and thus inhibits N leaching.

A disadvantage with this method is that the absence of precipitation and infiltration of water through the soil profile and the absence of a living root rhizosphere probably influences mineralization processes in the soil. Plant uptake of soil N can be estimated by sampling plants in plots without N application and analysing them for the N content. Nm during the growing season of a certain crop can be calculated as plant uptake of soil N plus residual mineral N in soil at the end of N uptake of the crop, minus soil mineral N in spring.

This method considers N mineralization not only in the topsoil but also in the subsoil as deep as the roots penetrate. It is assumed that Nm in unfertilised plots is equal to Nm in fertilised plots. This may not be quite correct, since priming effects from fertiliser N on mineralization have been observed. Another disadvantage with this method is that root N must be estimated, as root sampling is laborious and uncertain. A further disadvantage is that Nm is also affected by processes other than mineralization and immobilization, i.e. leaching, denitrification, ammonium fixation and release, atmospheric deposition and ammonia volatilization.

PLANT AVAILABLE WATER ESTIMATION

Plant available water (PAW) can be measured in the field or estimated with models. These usually rely on precipitation, evapotranspiration and soil water holding capacity. Maximum amount of PAW is dependent on soil water tension and rooting depth of the crop, which both are largely associated with texture. Several attempts have been made to interpret clay content and soil organic matter content into water holding capacity or PAW.

There are direct and indirect methods of measuring actual soil moisture content in the field and several ways of expressing the result quantitatively. The gravimetric method of measuring mass wetness consists of removing a sample by augering into the soil and then determining its moist and dry weights. This method is simple but destructive and time-consuming. Other methods estimate soil moisture by a calibrated relationship with some other measurable variable. These methods can be divided into volumetric methods measuring volumetric soil moisture and tensiometric methods, which give soil suction or water potential. Both quantities are related through the soil-water characteristic curve specific to each soil. Volumetric soil moisture content can be measured with an impedance measuring technique using an instrument that responds to changes in the apparent dielectric constant.

Site-specific fertilisation aims to match fertiliser dressings with crop requirements and soil nutrient contents as they vary across individual fields. Technology, such as global positioning systems (GPS), geographical information systems (GIS) and remote sensing, used for making rapid measurements of soil and crop properties, has made it possible to collect and analyze spatial data to make detailed yield and soil maps. Since fertiliser requirements are dependent on yield level and soil nutrient supply, such maps may constitute a basis for maps of fertilisation requirements.

CHAPTER 4

PHOSPHORUS FERTILISATION

Phosphorus (P) is an essential nutrient for all forms of terrestrial life and is one of the 17 chemical elements known to be required for plant growth. Agricultural soils generally contain from 800 to 2,000 pounds of total P per acre in the tillage layer. However, most of it is in insoluble compounds unavailable to plants. The remainder cycles within plants, animals, soil, and the soil solution in biologically available forms and organic P compounds. A simplified P cycle is depicted in Figure 1, showing the principal P inputs and sinks. In production agriculture, fertiliser and manure are the major P additions to this cycle. Without these inputs, intensive commercial agriculture would not be viable on many soils. However, proper management soils and P fertilisers is essential to protect water quality from degradation.

Water quality problems associated with phosphorus are generally confined to surface water. Phosphorus in most soils is tightly held to soil particles and does not leach. However, the P held in organic phases from residues such as manure can dissolve in water and be lost if improperly managed. Adsorbed P on soil particles can cause surface water contamination as P containing sediments move off the land in agricultural runoff. When large amounts of nutrients enter lakes and streams, they accelerate the natural aging process, or eutrophication, by enhancing the growth of algae and other aquatic weeds. As these plants

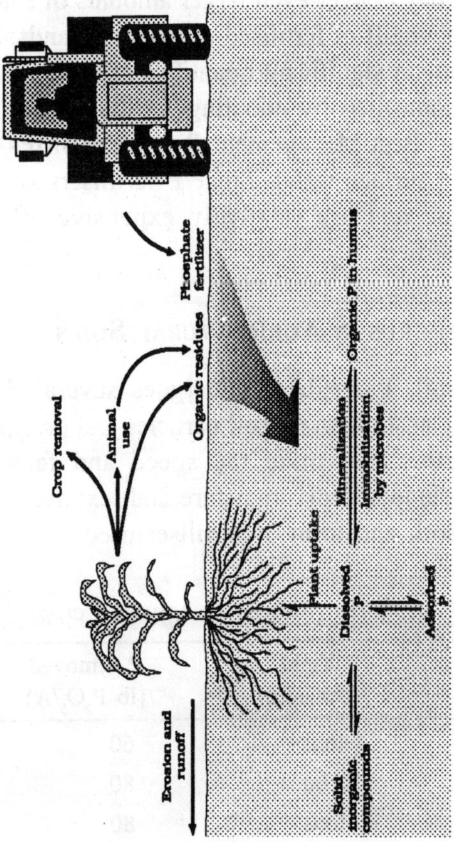

Figure 1: The phosphorus cycle in agricultural soil.

flourish, depleted oxygen and light reduce the survival of more desirable species and the natural food chain declines.

Eventually, impounded waters such as lakes, ponds, and reservoirs become overgrown with aquatic vegetation and, in a sense, die. A number of human activities contribute to P movement in our environment through runoff and erosion. Agricultural soils may deliver even larger amounts of P to water when sediments move off of fertilised or manured lands. Feedlot runoff also can contain significant quantities of nutrients. These P losses can be controlled by adopting fertiliser management and erosion control Best Management Practices (BMPs) which minimize runoff to surface water. Since P fertilisers are a non-renewable resource and are relatively expensive, efforts to conserve P are justified.

PHOSPHORUS IN AGRICULTURAL SOILS

When added to soil, P fertiliser undergoes several different reactions, including adsorption on soil particles and precipitation. A number of factors determine the speed and fate of the reactions. They include soil pH, moisture and texture, chemical properties of the soil, and form of fertiliser used.

Table 1: Phosphorus removed in harvested crops

Crop	Yield (per acre)	P removed (lb P_2O_5/A)
Alfalfa	4 tons	60
Corn (grain)	150 bu	80
Corn (silage)	30 tons	80
Barley	100 bu	55
Bromegrass/fescue	4 tons	65
Potatoes	350 cwt	55
Sugarbeets	25 tons	15

Sunflowers	2,000 lb	40
Wheat	40 bu	30

Phosphorus is found in both organic and inorganic forms in the soil. The organic forms are found in humus, manures, and crop residues and are important for supplying P to crops. These forms also tend to be more soluble and are subject to movement in the soil solution. Microorganisms break down organic matter in a process called mineralization, converting organic forms of P to plant available inorganic forms. The inorganic ions, $H_2PO_4^-$ and HPO_4^{-2}, are the primary forms of P taken up by plants. In general, the inorganic forms of P contained in soil minerals are relatively insoluble and are slowly available to plants. When making decisions regarding P fertiliser applications, it is important to account for the P available from the soil and from manures. Sewage sludge biosolids applied to cropland also contain significant quantities of P and should be credited appropriately.

MANAGING FERTILISER TO REDUCE PHOSPHORUS LOSSES

Applying P fertiliser at rates higher than production requirements is unwise from both environmental and economic viewpoints. Today, there is no agronomic justification for building P soil test levels higher than crop sufficiency levels. Phosphorus losses in surface runoff have been shown to increase with increased P application rates. Therefore, once the crop sufficiency levels have been reached in your fields, P applications should be made only as dictated by soil testing. Placement of P fertiliser will influence the amount of P available for transport to surface water.

Correct placement of fertilisers in the plant root zone will improve fertiliser use efficiency and seedling vigor, and reduce the amount of P in agricultural runoff. Phosphorus fertiliser should not be broadcast on the soil surface without incorporation, except on perennial forages. In established alfalfa stands, P fertiliser normally should not be applied in the late fall or winter when growth is minimal and runoff potential is high. Broadcast applications generally are less efficient and leave more P at the soil surface than banding (Figure 2). Band application at planting is considered the most efficient method for many crops. Subsurface placement is especially important under reduced tillage cropping systems to achieve maximum crop yields.

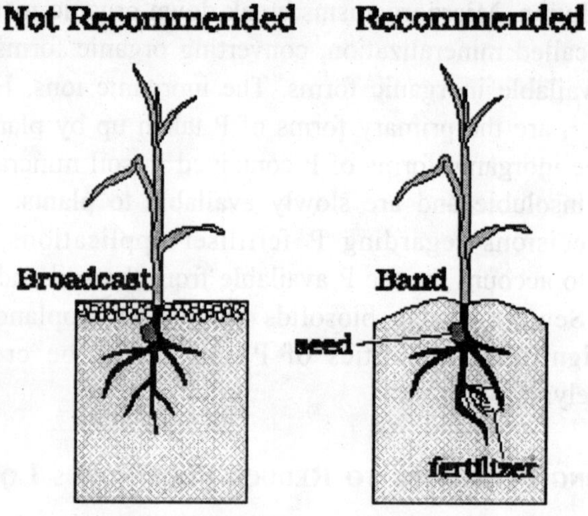

Figure 2: Phosphorus placement influences the amount available for transport. Band placement of P fertilisers is recommended for erosive soils.

Variable fertiliser rate management can improve both fertiliser use efficiency and economic returns. While this strategy can be adopted for any fertilised field, it makes the most sense in relatively large fields where the producer has knowledge of how crop yields and soil type vary across the field. To use a variable fertiliser rate strategy:

1. Divide the field into different management units based upon a map of yields or soil types.
2. Soil sample the management units separately.
3. Fertilise each unit according to P soil test level and yield capability.

Field maps should be modified at harvest as necessary to refine the boundaries of management units. Consult with your fertiliser dealer or crop advisor prior to adopting this BMP.

MANAGING MANURE TO REDUCE PHOSPHORUS LOSSES

Manure is an excellent source of P for crop production. However, if manure is not incorporated into the soil, runoff may carry both soluble and sediment-associated nutrients to surface waters. The most common strategies for manure utilization are (1) application for maximum nutrient efficiency and (2) application for maximum disposal rates of manure. While the second strategy presents a more difficult challenge from a water quality viewpoint, both management methods should consider application rates, timing, site characteristics, and water quality impacts. Manure managed for maximum nutrient efficiency is the most sound manure application program.

Producers need soil and manure analyses to determine the correct application rate based upon crop uptake of N and P. Either of these nutrients may limit application rate, as both nutrients are present in large quantities in manures. In many cases, the best program is to rotate fields receiving manures to avoid salt or nutrient buildup. As with commercial P fertilisers, manure should be incorporated immediately after application. Injection of liquid manure beneath the soil surface with specialized equipment is also a recommended practice. Unlike commercial fertiliser, the P content of manure can vary significantly. Approximate values are available for various manure sources (Table 2), but manure sampling and analysis are the best way to calculate nutrient credit.

Site characteristics such as land slope, surface residue, and proximity to surface water must be used to determine which management measures are needed to protect surface water from P enrichment from manure.

Table 2: Approximate P content of various manures when applied to land (wet weight basis)

Manure	% moisture	Average P_2O_5 content
Beef		
solid	48	14 lb/ton
liquid	91	27 lb/1,000 gal
Dairy		
solid	82	4 lb/ton
liquid	92	18 lb/1,000 gal
Swine		
solid	82	9 lb/ton
liquid	96	27 lb/1,000 gal
Sheep	72	10 lb/ton
Poultry		
without litter	55	48 lb/ton
with litter	25	45 lb/ton

In some cases, sites with excessive slope or highly erodible soils are not suitable for manure application, even with careful management. Application of manure on frozen or wet soils subject to runoff is not recommended. Runoff from feedlots and manure stockpile sites also can contribute nutrients to nearby surface waters. BMPs to reduce runoff impacts include use of yard shaping, settling basins, diversions, and filter strips (Table 3).

Table 3: Erosion control BMPs for reducing surface losses of phosphorus from crop fields

Best Management	Practice Description
Conservation tillage	Cropping system that maintains at least 30% of the soil surface covered with residues after planting
Conservation cover	Perennial vegetative cover established and maintained on highly erodible lands

	where other BMPs are insufficient to reduce adverse water quality impacts
Conservation cropping sequence	Crop rotation sequence designed to increase crop residues on the soil surface to reduce erosion
Delayed seed bed preparation	Cropping system in which all crop residues are maintained on the soil surface until three to four weeks prior to planting the succeeding crop
Grass filter strip	Permanent sod strip planted at the base of sloping fields or between the field and surface water bodies
Grassed waterway	Sodded channel that provides a non-erosive outlet for runoff
Contour farming	Crops planted on the natural contour of the land to reduce erosion
Strip cropping	Alternating strips of row crops and solid seeded crops planted on the contour
Terrace	Earthen embankment constructed across the slope to reduce slope length and runoff velocity
Diversion	Grassed channel constructed across the slope, uphill of a tilled field, to divert excess water to areas where it can be managed properly
Sediment control basin	Basins constructed to collect runoff and trap sediments
Constructed wetland	Artificial wetland created downhill from irrigated crop fields where sediment and runoff are collected and assimilated by growing vegetation

Owners of agricultural land should contact the Soil Conservation Service (SCS) for help in evaluating the erosion potential of their lands and in determining what control measures are needed. In some cases, the SCS has cost-share funds available to help producers install BMPs on their land. A number of management practices and structures for controlling runoff and erosion are currently available for use. In some cases, there is a trade-off between reducing runoff and increasing deep percolation to groundwater. BMPs for managing surface runoff and soil erosion are listed in Table 3.

Chapter 5

Organic Manures and Chemical Fertilisers

Integrated Plant Nutrition System (IPNM)

Integrated Plant Nutrition System is a holistic approach to plant nutrition by obtaining the nutrients from both inorganic and organic sources to maintain and sustain soil fertility and enhance crop productivity in a framework of an ecologically compatible, socially acceptable and economically viable situation.

Need for integrated use

- Organic manures sustain soil fertility at a low level of production.
- Chemical fertilisers have concentrated forms of nutrients. Application results in leaching, fixation and build-up of certain nutrients at the expense of others, resulting in nutrient imbalances.
- Fertiliser use efficiency (FUE) is low in all chemical fertilisers and organic manure when used singularly.
- Combined use of organic manures and chemical fertilisers increases FUE

Objectives of IPNS

— Increasing the fertiliser use efficiency.

— Increasing the return to investment on fertilisers.

— To use a balanced system of crop nutrition management.

— To account for the different amounts of plant nutrients, in a budget sheet giving the following:

1. Available in the soil.
2. Available from organic sources.
3. Available in crop residues.
4. Derived from biological fixation.
5. Derived from fertilisers and their residual effects.

When preparing a budget, a whole cropping system is looked into than a singular crop, over a long term.

Wrong in Fertiliser

— Before chemical fertilisers were introduced, organic forms were used.

— Green manures, crop residues and FYM were used in large quantities.

— A season of farming was followed by a fallow period for replenishment of soil fertility.

— Chemical fertilisers supplied sufficient plant nutrients for obtaining high yields. These fertilisers did not provide the crop requirement of all nutrients.

— The continuous use of chemical fertilisers destroyed the soil structure, turned the soil acidic and brought about nutrient imbalances resulting in nutrient interactions.

— The organic matter in the soil did not improve with application of chemical fertilisers resulting in erosion and loss of top soil.

— The yield increases obtained during the early years of chemical fertiliser use evened out and addition of extra fertilisers did not bring the desired benefits.

Farmer Practices of IPNS

— Vegetables, potato and green leafy vegetables are cultivated using both organic and chemical fertilisers.

— Here farmers obtain higher yields than farmers using either chemical fertilisers or organic manures.

— Extension of this practice to other crops has resulted in increases of yield and sustainability of production at a higher level.

Chemical Fertilisers

— To sustain crop yields, the rate of removal has to be balanced by added amounts. Use of chemical fertilisers is necessary for supplying the nutrient requirement but without recycling of crop residues, yields will suffer.

— If only organic manure is used land will benefit but yield will be lower.

— Availability of nutrients from organic manure is slow but long lasting.

— Use of organic manure improves the physical, chemical and biological conditions of the soil.

— Singular use of chemical fertilisers has an adverse effect on the soil structure. If the organic matter is low the bulk density increases.

— Use of chemical fertilisers along with organic manure gives a soil rich in nutrients with good physical and microbiological properties. This will increase the availability of nutrients.

— Application of organic manure results in formation of water stable aggregates which could resist erosion.

— Combined use of chemical and organic fertilisers increases cation retention and improves nutrient availability.

— Addition of manure widens the C:N ratio in soil. Singular use of chemical fertilisers narrows this ratio.

— Addition of mineral N fertilisers results in instant increases of N but organic manures increases the available N.

— Fixation of fertiliser P could be reduced and effectiveness of fertiliser K can be increased when chemical fertiliser is combined with organic manures.

— High analysis fertilisers have low contents of micronutrients, but combined use with organic manure makes these nutrients available to plants.

— Combined use of manure and chemical fertilisers results in higher return to investment and better cost-benefit ratios.

— Thus it is beneficial to use both manure and chemical fertilisers for crop production.

Organic Farming

— Organic farming systems do not use any synthetic form of plant nutrient.It also avoids the use of pesticides and weedicides. Organic manure of animal origin should not have artificial ingredients fed to animals and animals should not be treated with antibiotics or steroids as growth promoters.

— Obtaining green manure continuously from trees results in removal of plant nutrients from soils, which are not easily replenished. This results in poor growth of the green manure trees itself, which could result in a short fall of supply of the material. Thus a limitation to resource management occurs.

— Organic farming maintains the productivity at a low level. Where demand for food is becoming more increasing organic farming per se cannot meet the required demand.

— Plants obtain nutrient sources in the ionic form. Whether they come from organic or inorganic forms, the type of ions is the same.

— It is best to combine the use of both manure and mineral fertilisers rather than manage the singular use of manure.

Nutrient Balance

— IPNS attempts to keep a balance between crop removal and nutrient addition to the soil. In this respect a book keeping exercise has to begin with the first step of how much the soil already has.

— Soil analysis for evaluating the nutrient availability and its limitations that could reduce the productivity, has to be done at the begining.

— The yield goal of the cropping system of single or multiple crops that would be grown over time (say one year) has to be evaluated to arrive at the nutrient removal.

— The quantity of mineral fertilisers to be applied has to be decided on soil test values for P and K.

— The addition of organic manures to supplement the requirement of the minor and micronutrients has to be evaluated.

— Nitrogenous fertilisers to be used on crop demand rather than on blanket recommendations has to be practised.

— Limitations to FUE by drought, surface erosion, poor drainage are to be avoided.

— Correct application practices in basal and top dressing applications of both organic and chemical fertiliser have to be adopted.

— Crop residues have to be recycled to supplement the nutrient pool of the cropping system.

— If legumes are included in the cropping system, credit has to be given for possible biological nitrogen fixation (BNF).

Socio Economic Side

— Any IPNS practice has to be economically sound and socially acceptable. Organic manure should be available easily in acceptable forms and types.

— Crop residues are often burnt instead of recycling. Scarcity of labour, and the time gap between two crops may limit the use.

— Organic manure for many farmers is not available in their own holdings. The holding size often limits animal husbandry. Transported material is too expensive and transport to the farm is laborious.

— Chemical fertilisers are easily available and literature on their use is more comprehensive. Organic manure and crop residue management has not been given the same importance.

— Use of FYM, poultry litter and other animal wastes is popular among farmers. But sewage, sludge and urban compost are not popular among them.

— Financial credit is available only for mineral fertilisers. There is no organized credit system to help the farmers to use organic manure and crop residues.

— Chemical fertiliser market is controlled with a standard product but organic manure is sold without any standards.

— Storage and retail marketing of organic manure is limited. It is not easily accessible to most farmers.

— Farmers who have domesticated animals do not have facilities to collect,store and market the animal waste.

— The scope for IPNS is limited to a few high value crops in the present circumstances. Farmers growing these crops may be encouraged to follow IPNS. However, a large number who have access to organic manure do not use them.

IPNS and Environment

— Chemical fertilisers are considered to be environmental pollutants. Nitrogen fertilisers when used in excess can be leached or washed through erosion into waterways and reservoirs. Eutrophication of reservoirs have been reported from time to time.

— Nitrogeneous compounds in organic manures too are converted to nitrates. Any excessive usc and leaching could bring the same environmental hazard as using excessive amounts of urea or ammonium sulphate.

— Nitrogen fertilisers can acidify soils, but when used with organic manure can buffer acid formation to a certain degree.

— Liming can change the pH and improve CEC. Liming can be a corrective measure to improve the crop environment.

— Excessive build-up of P and K in soils results in nutrient interactions. In high P built-up soils, both zinc and copper availability can be a problem. At high levels of K in the soil, magnesium becomes limiting. By following IPNS these excesses can be cushioned.

— Excessive use of organic manures can limit the availability of copper as the latter is lost through leaching. Following IPNS will benefit the situation.

— Combined IPNS, benefits the development of soil structure and reduces erosion hazards.

— IPNS practices produce more healthy plants which minimizes the necessity to use pesticides and fungicides, which are harmful to the environment.

— Excessive nitrate leaching into groundwater and drinking water is harmful. IPNS can increase the anion retention when mineral and organic fertiliser is used together.
— Ploughing and land preparation is easy on lands following IPNS. Ploughing deep to expose the subsurface soil does not arise in these lands.

Fertiliser Efficiency

Chemical fertiliser in the form of salts, when added to soils gets converted into ionic forms after dissolving in the soil solution. Substances like urea under go extra cellular enzymatic decomposition to form ammonium compounds, which are either absorbed by the plant roots or converted to nitrates, which are absorbed or lost in leaching or converted to gases in the N cycle.

Organic manure produces organic acids (humic acid) which have the capability of holding cations and anions. Soils rich in organic matter can hold more cations than otherwise. Ammonium ion like other cations can be adsorbed to organic acid molecules after replacing H. This adsorption helps to release the cations slowly to the plants. If not for the organic molecules certain chemicals reactions can make most nutrients unavailable to the plants. Most plant nutrients are available at certain pH ranges. Addition of organic manures can buffer pH, making more materials available to plants. Soil organic matter has a high CEC values and CEC is pH dependent. If liming is done to increase pH then the CEC will also increase. The CEC from organic manure is temporary.

Application of Organic Manure

— Organic manures are of plant and animal origin.
— Plant materials have easily and slowly biodegradable materials. The variety, age and plant part determine the composition and the quickness of decomposition.

— Younger leaves and tender stems have more water, more N and less other minerals and decompos faster. Matured leaves and stems have more cellulose containing material, which takes a longer time to decompose.

— Crop residues such as straw have mature tissues with high C:N ratio. They take a longer time to decompose. Direct application will cause the temporary immobilization of soil N.

— Animal wastes have more N than plant parts. They decompose faster than plant materials. Application of N rich materials like blood meal will result in losses of N through volatilization.

— Composted material is suitable for direct application. But exposure of such materials to the elements results in nutrient losses through leaching.

— Care should be taken in using urban compost made out of sewage materials and industrial wastes. Such materials could contain both pathogenic microbes and heavy metals.

— Heavy metals such as cadmium, arsenic, lead and mercury are toxic to environment. These heavy metals are known to accumulate in the body.

— Organic manure decomposes faster in the soil. A crop residue takes one to two weeks to decompose depending on the moisture content of the soil and the type of material. Manure of animal origin decomposes within a few days.

— Application of organic manure should be timed in such a way to get the maximum benefits both in terms of plant nutrient supply and the addition of the carbon source.

— Incorporation to the soil is better than leaving at the surface. Some farmers give top dressings of welldecomposed manure, to crops.

— Poultry litter should not be applied immediately before seeding or transplanting as the ammonia gas released may increase the pH and kill the plants.

— Well-dried organic manure can be kept for a few months in polyvinyl bags.

— Liming should be done before adding organic manure in order to conserve N.

— Organic manure application to illdrained soils could result in the formation of toxic compounds; particularly if the material is rich in sulphur.

— Recent findings indicate that rice straw can be added to bog soils. The silica content of the straw benefits the soils.

CROPPING AND FEED PRODUCTION

All crops require adequate soil fertility for economical production. Manure is a valuable source of plant nutrients and organic materials including many living organisms. Improperly handled manure can be an environmental and health hazard to people and livestock. However, properly managed it is a valuable source of plant nutrients that can reduce the cost of crop production. When manure is applied to cropping areas, the application should occur before seeding or before there is significant growth of the crop that is to be harvested.

When bedding is used in livestock production, it along with unconsumed feeds will also be incorporated into manure. This should also be considered during manure application. Application in the springtime will generally ensure adequate time for soil assimilation. This will also generally reduce or eliminate contamination of new crops being grown as feed by manure that has been applied as a fertiliser to the land. Cultivation of cropland should be used to assist soil incorporation.

The application of manure should be handled to minimise the possibility of cross-contamination from manure to growing crops. Manure can be applied to land in the fall, but it should be incorporated into the soil to minimise leaching of nutrients during spring run-off. Some livestock operations may be in a position to utilise the manure output of other livestock enterprises or municipal sludge if it is properly handled. Manure or sludge from other sources can be an environmental or health hazard. Some knowledge of the manure or sludge is important.

It should not be from livestock or a source that has or may have diseases known to be transmitted to the livestock being produced or that will survive the rest period after application. There is the possibility that this manure could be contaminated with Salmonella and other pathogens unless the proper precautions are taken. When manure from other species or sludge from municipal sewage treatment facilities is used on land to raise livestock, it should be examined for the possible presence of industrial or other toxic chemicals.

Tillage: Tillage of the soil may be combined with the manure application to assist with soil incorporation.

Seeding: Treated seed should not be used as a feed or food unless this practice is allowed.

Growth: Herbicides and pesticides are to be applied at the rates and frequency as indicated by the instructions for use. Use all herbicides for the purposes indicated and store them in well-marked containers.

Harvest: Apply pre-harvest herbicides in accordance with the instructions.

Wintering: Grazing of crop residues is occasionally practised to extend the available feed supply.

PASTURE LANDS

— Native type

— Cultivated or established

— Improved pastures

Most normal practises of livestock production are conducive to good land stewardship. Good pasture management practises such as rotation and dragging to spread and disperse manure droppings will reduce possible cross-contamination between groups of animals. When manure is applied to pasture it should be timed so that an appropriate rest period can be given to the land.

This will allow destruction of bacteria that are potentially harmful to grazing livestock and that could be transmitted into the food chain/supply. Proper timing of the application of manure should insure that this is not a problem. Harrowing or dragging of the pastureland after spreading or grazing will break the larger clumps of manure into smaller pieces. This will encourage rapid drying, incorporation into the soil and subsequent destruction of bacteria.

Land Application of Manure

Manure Handling and Storage

The design of the livestock facility and the manurehandling component has to meet provincial and local regulations or by-laws for setback distances from adjacent properties and watercourses when applicable. It must be able to contain manure during runoff from a heavy snowmelt or other intensive runoff event. Its design must account for local soil and surface water conditions. Although it is difficult to prevent at certain times of the year, it is desirable to handle water run-off from contaminated feeding areas so as to minimise its consumption by livestock.

Manure Application Rates

The key to using manure successfully is proper management. Used properly, manure will promote good crop utilisation and soil improvement while minimising any hazard to the environment. Apply manure at rates and times of the year that:

— are compatible with the nutrient requirements and the growing characteristics of the crop;

— are compatible with soil characteristics, drainage and the slope of the land; and

— protect the quality of surface and ground water.

Depending on the provincial or local regulations, manure application rates should be based on both the nitrogen and phosphorus requirements of the crop being raised on the land. If these nutrients are applied in excess, the excess may be lost to the environment where it could cause a problem. For example, applying manure to the required nitrogen level could lead to excess phosphorus and in turn lead to environmental contamination. Nutrient requirements for a crop depends on several factors such as:

— plant requirements;

— amount of plant nutrients already in the soil;

— amount of manure applied in previous years;

— length of time manure is left on the soil surface before it is incorporated;

— amount of crop residue present in the field; and

— the presence or absence of a winter cover crop.

Adequate Land Base

Enough land should be available to receive the amount of manure that is produced each year. Such land can be either owned or available through land spreading agreements with neighbours. Manure application rates should be based on soil tests and the requirements of the crop to be planted or the requirements of existing forage or pasture crops. Consult with an agricultural engineer or professional agrologist to calculate the area of land and the application rate that is required when spreading manure.

Incorporation of Manure

Rapid incorporation of the manure into the soil by injection, plough-down or other tillage methods has several benefits:

— maximises nutrient uptake by crops by minimising ammonia losses;

— reduces risk of manure runoff; and

— minimises the generation of odours.

Environmental Contamination

To minimise contamination to soil and ground water consider the following in a manure management plan:

— soil characteristics;

— drainage;

— compaction potential with heavy equipment;

— level of nitrates in the root zone;

— soil cover at application time; and

— expected spring runoff.

In all cases follow provincial or local regulations, bylaws and codes of practice that are in place for the spreading of manure. To prevent contamination of runoff water, manure should not be spread in the following areas or conditions:

— on steep slopes where erosion and surface water runoff is likely to occur;

— on saturated soils, where manure will not infiltrate into the soil;

— within the high water mark of field depressions during the times of the year when there is a high risk of direct surface runoff to a natural watercourse;

— during high winds when drifts can occur or when using irrigation or aerial spraying equipment;

— on frozen or snow covered ground when the runoff during snow melt could be contaminated; and

— on areas where there is standing water.

Many materials that are organic in nature are suitable for IPNS. Materials that can be used are classified into:

i.	Plant residues	-	Crop residues Green manure
ii.	Animal wastes	-	Animal manure Slaughter house wastes.
iii.	Compost	-	Consists of both plant and animal materials whichare wastes which also includes city garbage.

Crop residues

Rice straw, corn stover and vegetable residues can be applied directly or after composting. Direct application is cumbersome and due to high C:N ratio of the materials, it takes a longer time to decompose in the soil.

Green manure

Are of two types

— Which are grown in situ and ploughed into the soil.

— Lopped leaves and tender stems which are applied to the soil.

Grown insitu

Plants like Crotalaria juncea and Sesbania rostrata can be grown before establishing crops and ploughed in at flowering during land preparation. Soils with low content soil organic matter will lose more carbon if N high green manure is used.

Animal wasıse

— Animal dung and urine of large animals found mixed with

bedding and feed stuff are used as farm yard manure (FYM). Poultry litter can be used both as a basal and top dressing. The material content of the FYM and litter varies considerably depending on many factors.

— Slaughter house wastes and fish meal are not available in Sri Lanka in large amounts due to poor collecting facilities. If collected and treated properly it could become and important source of organic manure.

City wastes

City garbage can be composted and used. It is high in extraneous materials like silica but is a useful material if properly composted. It can be used for floriculture and fruit culture in places close to cities. Above materials are available for IPNS. The quantities that may be ideal for application would be comparatively larger than the rates of chemical fertiliser. But it is important to add any available amount of organic material available at appropriate times.

CHAPTER 6

CROP ROTATION AND FERTILISATION

The flow of nutrients in and out of agricultural systems is usually much greater, with lower nutrient storage capacity and less nutrient cycling efficiency. The sustainability of agricultural systems is therefore greatly dependent on optimizing the balance between inputs and outputs of nutrients. One way of achieving this balance is by ensuring that nutrients removed from the soil are returned to the system. In this manner, a nutrient cycling mechanism that is essential for the sustainability of the production system can be established.

In any agricultural system, the harvested produce is the major avenue of nutrient removal from the soil. For instance, a cereal crop like wheat can, in the grain alone, remove up to 200 kg N and 40 kg P ha^{-1} $season^{-1}$. This removal can greatly offset nutrient cycling and affect the sustainability of the agricultural system, thus necessitating substitution of the nutrients exported with crops from the field. Conventionally, these nutrients are replenished by promoting such practices as legume-based crop rotation or applying organic and inorganic fertilisers.

Legume-based crop rotation is an effective and often profitable way of supplying N and improving soil properties. Organic and inorganic fertilisers are also important not only in elevating the nutrient status of the soil and stimulating microbial activity and plant growth but, by altering the pools and fluxes of C and other chemical elements, may bring about changes in the chemical and microbial properties of soils.

CROP ROTATION AND ITS EFFECT ON SOIL

Crop rotation - a system of the repetitive cultivation of a sequence of crops in an orderly and recurrent succession on the same land- has been practiced for more than 3000 years and is known to many cultures and civilizations. Its beneficial effect in maintaining productivity has been recognized throughout history although the underlying reasons behind such effect were unknown until the recent past. For instance, it was not understood why crops grown in rotation often yielded more than those in monoculture until the later part of the 19th century when the ability of legumes to fix atmospheric nitrogen was uncovered.

As a result of this discovery and the coming to light of other beneficial effects of crop rotation in subsequent years, the practice became more popular and its use increased. However, with the innovation and development of modern farming techniques and price support systems in the 1950s through late 1970s, specialized cropping increased dramatically at the expense of crop rotation. Increased availability of chemical inputs such as nitrogen fertilisers, herbicides and pesticides, resulted in diminished popularity and use of crop rotation as did the introduction of improved crop varieties, mechanization, marketing practices and policies.

In the 1980s, renewed interest in crop rotation evolved due mainly to concerns over the impact of agriculture on environmental quality, decreased commodity prices, and the need to increase farm profitability through diversification. As a result of its beneficial effects in reducing chemical inputs and improving soil properties, crop rotation has now received significant attention and its role in environmental protection and sustainable agriculture is ever increasing.

Soil Chemical Properties

Crop rotation is known to beneficially influence many soil chemical properties including organic C content, N supply and transformation, pH and amount and availability of P, K, Ca and Mg. Crop rotation generally results in increased organic matter content compared to continuous cropping or fallow. This was demonstrated on the Morrow plots in Illinois where more than a century of research revealed a significant increase in soil organic matter content under crop rotation. Organic matter contents were highest in a maize (Zea mays L.)-oat (Avena sativa L.)-hay rotation, intermediate in a maize-oat rotation and lowest in a continuous maize rotation (Figure.1).

Similarly, in several long-term experiments (some dating from the early 1900s) at many localities in semi-arid regions of the US Pacific Northwest, greater increases in soil organic matter content of rotation soils compared to continuously cropped or fallowed soils were reported by Monreal et al. The effect of crop rotation on soil organic matter content is greatly dependent on rotation length and the inclusion of legumes in the rotation sequence. Crop rotations for periods of four or more years generally increase soil organic matter content while shorter rotations tend to result in organic matter levels slightly higher than continuous cropping.

This has been shown on the Breton plots in Alberta where a 60-yr study on different soils showed the organic matter of a 5-yr rotation to be higher than that of a 2-yr rotation. In another long-term experiment at Saskatchewan, Campbell et al., investigating the effects of crop rotations and various cultural practices on the quantity and quality of soil organic matter, found increased organic matter content with increasing frequency of cropping and the inclusion of legumes as green manure or hay crop in the rotation. They observed the organic matter content to be directly related to cropping frequency and found that the inclusion of sweet clover (Melilotus officinalis L.) green manure or lucerne (Medicago sativa L.) hay in cereal rotations increased

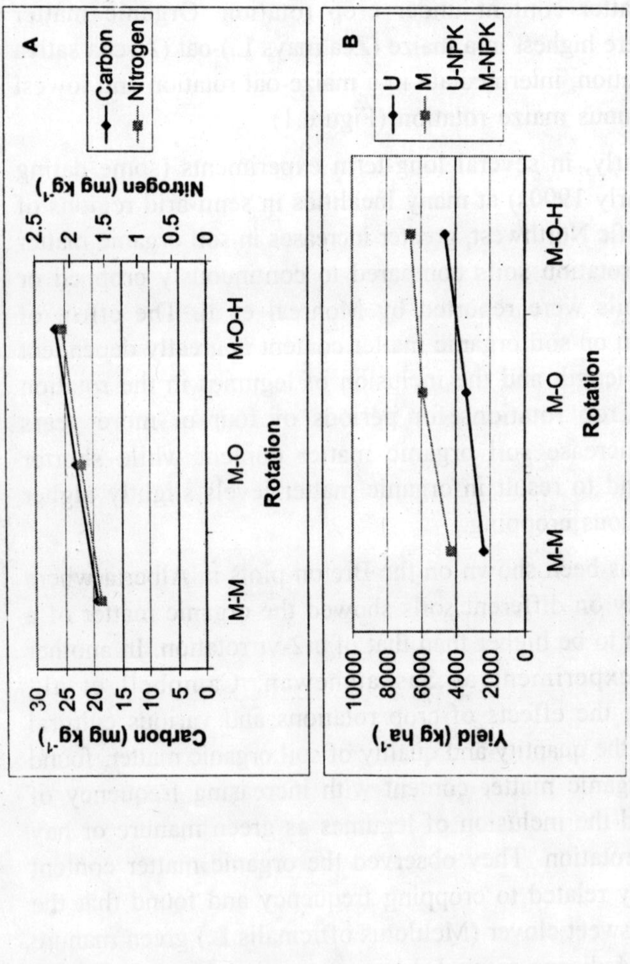

Figure 1: Mean soil carbon and nitrogen contents (A) and maize yield (B) in the Morrow Plots in relation to rotation (M-M = maize-maize; M-O = maizeoat; M-O-H = maize-oat-hay) and fertilization (U=unfertilized; M=manure; UNPK= NPK only; M-NPK=manure+NPK). From Aref & Wander.

soil organic C content compared to 2-year and 3-year fallow rotations.

Likewise, Janzen, showed that a 6-yr hay-containing rotation had slightly higher organic carbon content than a 3-yr green-manure-containing rotation. These differences in organic matter content between the longer and shorter rotations are due primarily to differences in the amount of biomass production. It may also be due to higher inputs of residue N relative to residue C which could have resulted in the conservation of C during decomposition. Soil N content under crop rotation is increased in two major ways: N input from legume crops in the rotation and reduced leaching of N.

On the Morrow plots in Illinois, soil N content in the maize-oat-hay rotation was significantly higher than in the maize-oat rotation and continuous maize cropping (Figure1). The increase was attributed to the inclusion of legumes in the rotations. Contributions of N from legumes in rotation are well-documented. As much as 10% of the rotation effect on grain yield of a subsequent cereal may be due to N benefits from the legume crop in the rotation. Ebelhar et al. and Hargrove estimated that the contribution of legume cover crops as source of N to subsequent non-leguminous crops may be as high as 120 kg N ha^{-1}.

McVay et al. studied the effects of winter legume cover crops on soil properties and nitrogen fertiliser requirements at two localities in the Coastal Plain region of Georgia (USA) and concluded that an adapted winter legume cover crop can replace all the fertiliser N required for optimal rain-fed grain sorghum (Sorghum bicolor (L.) Moench) and up to two-thirds of that required for maize productions. Leaching of NO_3-N under crop rotation is reported to be considerably lower than under continuous cropping or fallow.

In two separate long-term crop rotation experiments in Saskatchewan, deep-core soil sampling and analysis showed that NO_3-N leaching from legume-containing rotations was significantly lower than either fallow or continuous cropping. Other studies also observed lower NO_3-N leaching under rotation than under continuous cropping. A reduction in NO_3-N leaching is essential not only to maintain the available nitrogen level in the soil but also to reduce potential environmental contamination.

With increases in organic matter returns and availability of N, microbial activity can be stimulated thereby enhancing the decomposition process and release of P, K, Ca, Mg and various micronutrients. Inclusion of legumes in the rotation may be particularly important in facilitating the decomposition and nutrient release processes by reducing the C:N ratio of organic materials returning to the soil.

Additionally, legumes in rotation can increase available P levels by altering the pH of the rhizosphere and exchangeable K level through relocation of the ion to the soil surface from deeper in the soil profile. Increased availability of micronutrients, including Fe, Cu, and Zn because of microbiologically enhanced chelation may also be a beneficial effect of legume-based crop rotation.

Soil Microbial Properties

Crop rotation generally supports greater soil microbial biomass and numbers than monoculture. The difference is often related to changes in C input and may vary depending on crop type, frequency of cropping and rotation length. This has been demonstrated in several long-term experiments conducted for 28 to 64 years in major ecological zones of the Canadian prairies as well as in the midwest and Pacific northwest regions of the USA.

Collins et al. examined the inclusion of summer fallow and a legume crop in a long-term rotation experiment ongoing since 1931 at Pendleton, Oregon. They reported that soil microbial biomass C and N as well as populations of bacteria, fungi and actinomycetes were significantly higher under monoculture wheat than under wheatfallow rotation. Further increases occurred as a result of the inclusion of a legume crop in the system.

Microbial biomass and numbers under wheat-pea (Pisum sativum L.) were higher than under wheat-wheat or wheat-fallow. The Appalachian Piedmont near Raleigh, North Carolina, Kirchner et al. also found soil microbial biomass C and N, Bacillus spp., actinomycetes and total culturable bacteria to be significantly higher in soil with maize-sweet clover rotation than with continuous maize. The differences in soil microbial biomass and numbers due to the inclusion of fallow and a legume crop in a rotation are closely related to the amount of organic C returned to the soil.

This was demonstrated by Campbell et al. in a long-term crop rotation experiment at Indian Head, Saskatchewan. Variables examined in the investigation included cropping frequency, green manuring and the inclusion of a grass-legume hay crop in a predominantly spring wheat production system. It was noted that soil organic C and microbial biomass C and N increased with increasing frequency of cropping and with the inclusion of the legume as green manure or hay crop in the rotation.

Microbial biomass C and N were positively correlated with the amount of crop residues (including roots) returned to the soil over the period of the study. Similarly, a significant decrease in soil microbial biomass and numbers as a result of the inclusion of summer fallow in a rotation was observed in a separate long-term experiment at Swift Current, Saskatchewan, and attributed to differences in carbonaceous residues returned to the soil.

The effect of rotation length on microbial properties of soil was investigated by McGill et al. in the long-term Breton Plots in Alberta. A two-year fallow-wheat rotation was compared with a five-year rotation consisting of spring wheat-oat-barley (Hordeum vulgare L.)-forage-forage. The amount and turnover rates of microbial biomass in the five-year rotation were significantly higher than in the two-year rotation. The five-year rotation contained 117% more microbial biomass than the two-year rotation. It was concluded that the longer rotation practice exerted a greater effect on the amount of organic matter by controlling both input of C and biomass turnover.

Martyniuk & Wagner found higher numbers of bacteria and actinomycetes in a rotation including maize, oats, wheat and red clover (Trifolium pratense L.) than in either continuous maize or continuous wheat. The rotation soil contained 2.1% organic matter, a significantly higher level than that of the maize or wheat soils, which contained 1.3 and 1.5% organic matter, respectively. The difference in microbial populations was therefore attributed to differences in organic substrate effects that were brought about by the rotation length and the inclusion of red clover in the rotation sequence and the N fixed by the legume.

The residues from red clover and from subsequent crops that benefited from the fixed N were presumed to have provided more organic substrate for the microorganisms. Unlike frequent annual cropping which reduces loss of soil organic C, fallow in rotation is known to accelerate C loss. A decrease in microbial biomass and numbers by the inclusion of fallow in crop rotation is therefore because of the negative impact of fallow on the C content of the soil. On the other hand, the inclusion of legumes in a crop rotation and prolonging the length of the rotation are advantageous since they enhance and/or maintain the organic C level, thereby impacting favourably on microbial properties of the soil.

CHEMICAL AND MICROBIAL PROPERTIES OF SOIL

Fertilisers are defined as organic or inorganic materials of natural or synthetic origin which are added to soil to supply certain elements essential to the growth of plants. Two such materials that are commonly used and routinely applied to soils are manure and inorganic NPK fertilisers. Manure refers to faeces and urine excreted by cattle, dairy, poultry or other domestic livestock, including the bedding material. Use of manure for crop production is an ancient practice and an important way of recycling nutrients.

Though the composition and amount of nutrients in manure are variable depending on the type of animal, ration fed, amount and type of bedding material, collection system, and management between production and use, manure contains all the essential chemical elements needed by plants and its potential nutrient contribution is quite considerable. The use of inorganic fertilisers began with the development of the synthetic fertiliser industry in the 19th century and vastly expanded during the post-world war II era.

It has since been an integral part of modern agriculture and a major contributing factor towards the dramatic increase in food production in the last century. Inorganic fertilisers are now the most widely used sources for supplying nutrients to crop plants and have become an economic necessity in many production systems. Unlike manures which are composed of much C and relatively small percentages of several plant nutrients, inorganic fertilisers contain one or two elements in a proportion required for the production of high yields of crops.

The bulk of inorganic fertilisers consist of materials containing the three principal plant nutrients N, P and/or K in concentrations much higher than in manure. For instance, the concentrations of N, P and K in commonly used fertiliser materials such as urea, superphosphate and potassium chloride are, on average, 44, 15 and 45%, respectively, and these levels are 15-20 times higher than in manure. When manure and NPK fertilisers are applied to the soil, a complex series of chemical and microbial reactions may take place that not only influence plant growth and development but also cause various short- and long-term changes in the chemical and microbial properties of the soil.

Effect of Manure

Soil chemical properties

The bulk of the solid matter in manure is composed of organic compounds and partially degraded plant material. Of the total dry matter content of manure, which ranges from 420-500 g kg^{-1}, C accounts for about 150 g kg^{-1} (30%). Manure addition to soils can therefore result in substantial C inputs and significant increases in organic C levels of soils. In the classical experiment at Hoosfield (Rothamsted), annual application of manure at a rate of 35 tonne ha^{-1} to continuous spring barley for over 140 years resulted in an exponential increase in soil organic C levels of about three-fold over that in the unfertilised soil (Figure 2).

Under tropical conditions in the northwest of India, yearly applications of 45 tonne ha^{-1} manure increased the soil organic C level from an initial value of 0.47% in 1967 to 1.8% in 1995. In the experiment at Hoosfield, manured plots contained about 50% higher organic C than unfertilised plots 104 years after manure addition had been discontinued, thus showing the stability and persistence for many years of organic material derived from manure. Several other researchers have observed similar soil organic C increases and substantial residual effects as a result of manure additions.

Manure is well known as a valuable source of N, P, K and other nutrients. On average, manure contains 2.0, 0.5 and 1.5 % (on dry-mass basis) of N, P and K, respectively, and significant amounts of Ca, Mg, Na and many trace elements. When added to the soil, manure can furnish the nutrients it contains and elevate soil nutrient levels. Significant increases in total N and NH_{4-} and NO_{3-} N with increasing rates and number of annual manure applications have been reported.

Manure also increases the availability, persistence and movement of P and results in a substantial build-up of exchangeable K, Na and Mg. With a single manure application,

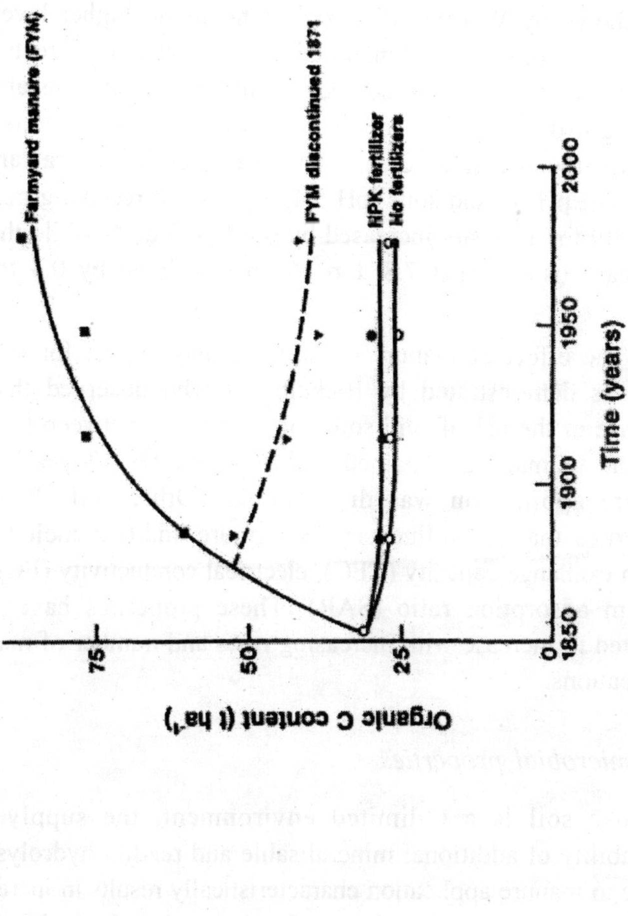

Figure 2: Changes in soil organic C content in the Hoosefield continuous barley experiment at Rothamsted.

Eghball & Power found an increase from 81 to 114 mg kg^{-1} in available P levels. In the exhaustion land trials at Rothamsted, P and K residues from previous manure application had remained in forms available to plants for more than 50 years. Similarly, in the Hoosfield continuous barley experiment, P and K residues from 20 applications of manure were still available to plants after 100 years.

The significance of manure as a source of micronutrients was shown by Warman & Cooper who found higher levels of B, Cu, Fe, Mn and Zn in manured than in unmanured treatments after 3 years of fresh and composted manure applications at rates ranging from 5 to 10 tonne ha^{-1}. Manure amendments can increase the pH of acid soils and decrease that of calcareous soils. The pH of acid soils (pH 5.4, top 15 cm) receiving manure annually for 11 years increased by 0.4 to 0.5 units while that of a calcareous soil (pH 7.8, top 15 cm) declined by 0.3 to 0.7 units.

The effect of manure on soil pH may persist for several years as demonstrated by Bickelhaupt who observed that an increase in the pH of acid soil from pH 5.7 to between 6.7 and 7.3 due to manure remained undiminished for 12 years after manure application was discontinued. Other soil chemical properties that are influenced by manure addition include the cation exchange capacity (CEC), electrical conductivity (EC) and sodium adsorption ratio (SAR). These properties have been reported to increase with increasing rates and number of manure applications.

Soil microbial properties

Because soil is a C-limited environment, the supply and availability of additional mineralisable and readily hydrolysable C due to manure application characteristically results in increased biomass, activity and abundance of soil microbes. In the Northern Guinea savanna zone of Nigeria, microbial biomass C of a Typic

Haplustalf manured for 45 years with cow dung showed an increase of more than 76% over the unmanured control.

Conducted under both tropical and temperate conditions, have also shown significant increases in microbial biomass C, N and P in soils receiving manure. The positive impact of manure addition on the density of microbes was demonstrated in the long-term field plots at Sanborn. Soils from treatments that received manure had higher numbers of bacteria and actinomycetes than those that did not.

NPK Fertilisers

Soil chemical properties

The purpose of inorganic fertiliser applications generally is to increase soil fertility and enhance crop production. The resultant enhanced plant productivity and associated greater returns of organic material in the form of decaying roots, litter and crop residues often give rise to higher soil organic C levels. As illustrated in Figure 2, long-term NPK fertilisers on a yearly basis to the Hoosfield continuous barley experiment resulted in an organic C level that was 15% higher than in the unfertilised treatment. Similar increases in soil organic C levels due to long-term NPK applications have been observed in other studies .

There are, however, reports of decline or no change in organic C content of soils receiving prolonged applications of NPK fertilisers. For example, in the long-term Morrow plots, Aref & Wander observed significant soil C losses throughout the history of the trial as a result of continued NPK applications. In Ranchi, India, annual applications for over 20 years of 100, 60 and 40 kg ha^{-1} of N, P_2O_5 and K_2O, respectively, did not cause any change in the organic C level of the soil.

It appears, therefore, that information on the impact of prolonged NPK applications on soil organic C content is inconsistent, probably due to influences of crop, soil and environmental factors. Application of NPK fertilisers can result in a substantial build-up of the total and readily available levels of soil nutrients. Shen et al. reported that a soil at the Broadbalk long-term experiment which had received 144 kg ha^{-1} fertiliser N each year for 137 years contained 3600 kg ha^{-1} organic N in the top 23 cm, compared to 2900 kg ha^{-1} in the soil receiving no N.

A long-term soil fertility monitoring programme at 11 major agroclimatic regions in India revealed that the availability and concentration of readily soluble P and K were much higher where these nutrients were applied. Similar increases in total and extractable N, P and K contents were observed in another long-term experiment at the Research Institute for Cereals and Industrial Crops, Fundulea, Romania. After 26 years of annual application of 160 kg ha^{-1} fertiliser N, total N in the fertilised soil was 13% higher than in the unfertilised control.

After the same period, plots receiving 160 kg P2O5 ha^{-1} yr^{-1} contained 126 mg kg^{-1} readily soluble P compared to 10 mg kg^{-1} P in unfertilised plots. Exchangeable K in the unfertilised plots declined from an initial level of 234 to 218 mg kg^{-1}, while it increased to 327 mg kg^{-1} in plots that had been receiving 120 kg K_2O annually. Nutrient balances calculated for several of the Rothamsted long-term experiments have demonstrated considerable carry-over effects of continued NPK applications. It was shown that up to 85% of the P and 40% of the K added as inorganic fertilisers over the last 100 years or so had been retained in the soil and were plant-available.

The increase in organic N, however, was equivalent to only 3% of the fertiliser N applied over a period of more than a century, thus showing that the net retention and residual value of fertiliser N is relatively small . Most N-containing inorganic fertilisers, unless specially treated, tend to acidify soil. This is mainly due to the impact of certain carriers that supply ammonia or result in its production when added to the soil.

Upon oxidation, the ammonia compounds can release H^+ ions which are potential sources of soil acidity. Several reports refer to declines in soil pH as a result of continued use of N-containing inorganic fertilisers. Prasad & Goswami found that continuous application for 18 years of N as ammonium sulphate, whether or not applied along with P and K fertilisers, decreased the pH of a soil from 5.5 to 3.9.

A decrease of 0.4 pH units in the soil at the Fundulea long-term experiment described above occurred as a result of annual application of 160 kg ha^{-1} ammonium nitrate. Such soil pH declines as well as decreased nitrification and nutrient mineralization that may result from the use of nitrogenous fertilisers could be unfavourable for soil microbial growth and activity.

Soil microbial properties

The effects of inorganic NPK fertilisers on soil microbial biomass and numbers are inconsistent. In some cases microbial biomass and numbers were stimulated while in other cases there were the opposite or no effects. In the long-term experiment in Saskatchewan, significant increases in soil microbial biomass and numbers due to fertilisation were reported. Studies in India and Japan also found similar responses to continuous applications of NPK fertilisers. The increases were attributed to improvements in soil fertility and the associated enhanced plant growth and higher rhizodeposition.

By contrast, McGill et al., studying the dynamics of soil microbial biomass C and N in the long-term Breton Plots, observed decreased microbial biomass due to NPK application. Experiments conducted for more than 10 years at two sites in Kansas also showed that soil microbial biomass C and N of plots fertilised with 224 kg N ha^{-1} yr^{-1} were significantly lower than that of the unfertilised control plot. In a study at the Saskatchewan long-term experiment, the amount of soil microbial biomass C in plots that received continuous NPK fertilisers was not significantly different from that in the unfertilised control.

The lack of response or decrease in microbial biomass and activity could be due to acidifying effects of the fertilisers and the resultant impairment of the survival capacity of many of the soil microorganisms. Generally speaking, information on the impact of NPK fertilisers on soil microbial biomass and numbers, in addition to being conflicting, is also very limited and more research in this regard is needed.

The interaction of crop rotation and fertilisation practices in influencing soil chemical and microbial properties is often positive. Application of organic and/or inorganic fertilisers to crops in rotation results ins increased plant biomass and hence greater return of organic material and increased soil organic C content. In addition to its nutritional value and enhancement of plant biomass, manure addition to crops in rotation is especially important in directly contributing to the organic matter of the soil by returning organic C.

Soil nutrient levels are also elevated when manure and inorganic fertilisers are applied to crops in rotation due to both the favourable impact of crop rotation as well as additional inputs from the fertilisers. In a rotation system consisting of finger millet (Eleusine covacana (L.) Gaertn.)-cowpea (Vigna unguiculata (L.) Walp.)-maize in Tamil Nadu, India, the available N status of the soil was maintained at a high level of 240 kg ha^{-1} with a combination of manure plus NPK fertilisers compared to 154 kg N ha^{-1} without fertilisation during 14 years of continuous cultivation.

The available P content in the fertilised treatment was 21 kg ha^{-1} against 5 kg ha^{-1} in the unfertilised treatment. Prasad & Goswami observed that the available K content of a soil cropped to maize-wheat rotation with application of manure and NPK fertiliser annually for over 28 years was higher by about 20 mg kg^{-1} than the unfertilised rotation soil. These data are consistent with those of other long-term experiments.

In the long-term Breton Plots, soil microbial biomass C and N of soils cropped to wheat -oats-barleyforage-forage rotations were 320 and 21 mg kg^{-1} without fertilisation and 390 and 41 mg kg^{-1} with manure application, respectively. In a two year rotation consisting of maize, soybean (Glycine max. (L.) Merr.) and barley, Kanazawa et al. observed that biomass C and populations of bacteria, actinomycetes and fungi in soil that received chemical fertilisers and manure, either separately or together, were significantly higher than in soil receiving no fertiliser.

Many other studies have indicated similar responses to the combined use of crop rotation and fertilisation practices. With improvements in soil chemical and microbial properties, it should be expected that productivity of the soil would be enhanced, resulting in substantial crop yield increases. Yields of crops from rotations, which are often higher than those in monoculture, have been found to be further enhanced by manure and NPK fertiliser applications.

In the long-term Morrow Plots, for example, yields of continuous maize plots have always been much lower than yields of maize-oat and maize-oat-hay rotations, and fertilisation with manure and/or NPK resulted in greater yields than unfertilised rotations (Figure 1). In two long-term experiments that have been ongoing since 1923 at Skierniewice (Poland), yields of several field and vegetable crops were highest in legume-based crop rotation with manure, considerably lower in rotation without legumes or manure and lowest in unfertilised monoculture. Other long-term studies have also shown similar yield increases and favourable influences of crop rotation and fertilisation.

The chemical and microbial quality of a soil and productivity of crops can be enhanced by crop rotation, manure addition and appropriate use of NPK fertilisers. The positive interaction and additional benefits accrued by crop rotation and fertiliser application emphasize the importance of integrated crop and nutrient management systems to the maintenance and improvement of the quality and productive capacity of soils.

CHAPTER 7

FERTILISERS FOR GREENHOUSE PRODUCTION

Fertilisers are salts. Salts are chemical compounds that contain one positively charged ion (cation) bonded to one negatively charged ion (anion). When a salt is placed into water, the two ions separate and dissolve. An example of a fertiliser salt is calcium nitrate, which contains one calcium cation and a nitrate anion. Other examples include: ammonium phosphate, magnesium sulfate, potassium nitrate and ammonium nitrate. Fertiliser concentration of a solution can be determined by measuring the ability of a solution to conduct an electrical signal.

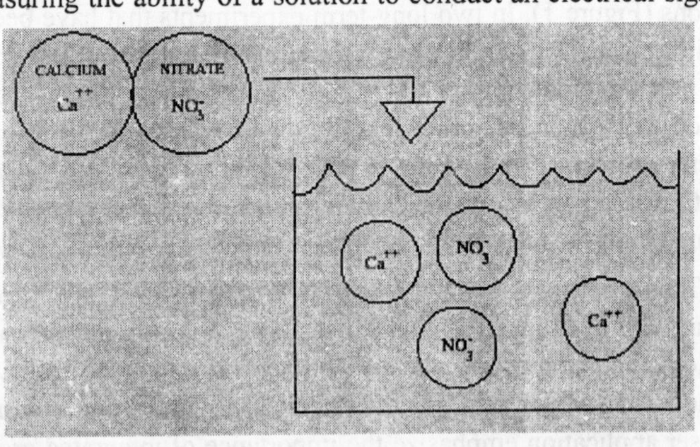

Electrical conductivity meters, often called soluble salts meters, measure the concentration of salts/ions in solution; therefore, a grower can always measure the amount of fertiliser being applied to a crop. However, electrical conductivity meters do not specifically measure which specific salts are in solution. For example, an electrical conductivity meter can not tell the difference between table salt, which is dangerous to plants, and potassium nitrate, which is useful for plants. Ions dissolved in water are taken up through the roots and distributed within the plant.

Plants actually expend energy to take up most ions, however, calcium is thought to only come along for the ride, i.e., plants don't actively take up calcium, it just comes into the root with the water. Once inside the plant, ions are recombined into compounds useful for plant growth. The most common cxample of plant metabolism involves photosynthesis, during which water is combined with carbon dioxide to form starch or sugars. Another example is the chlorophyll molecule shown below that contains:

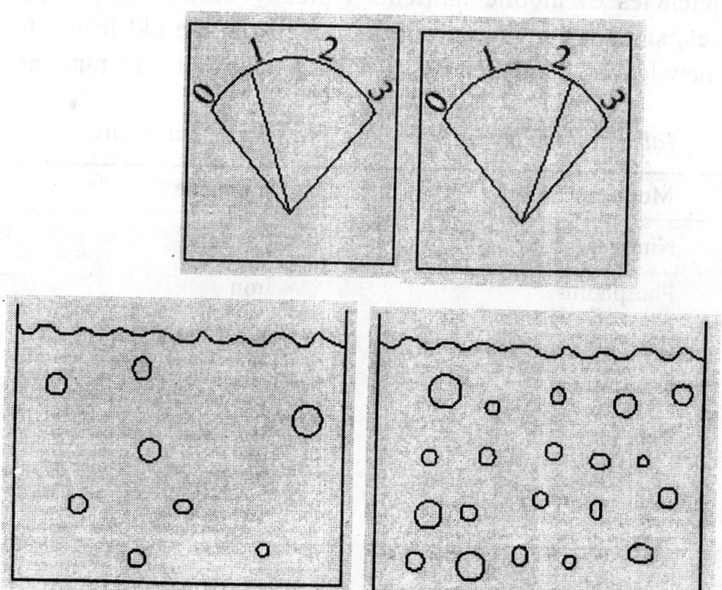

— 55 carbon atoms
— 60 hydrogen atoms
— 5 oxygen atoms
— 4 nitrogen atoms
— 1 magnesium atom

Therefore, for the plant to build one chlorophyll molecule, the leaves must take in carbon dioxide, for the carbon and oxygen; the roots must take in water, for the hydrogen and oxygen; and the roots must take up nitrogen and magnesium provided from the fertiliser applied. Once inside the plant, some nutrients can be mobilized to support new growing tissues, while other nutrients are fixed in older plant tissues. This fact helps to diagnose some plant nutrient deficiencies.

For example, if a plant is deficient in an immobile nutrient, then deficiency symptoms occur in the new growth, since the older tissues "hold" on to the immobile nutrients. In contrast, deficiencies of mobile nutrients typically occur in the older leaves, since the mobile nutrients move from the old leaves to the new leaves. Plants require different amounts of each nutrient.

Table 1: Mobility of individual nutrients within plants.

Mobile	Immobile
Nitrogen	Calcium
Phosphorus	Iron
Potassium	Manganese
Magnesium	Zinc
Sulfur	Copper
	Boron
	Molybdenum

Carbon, hydrogen and oxygen are required in the greatest amounts; however, these are taken up by the plant in the form of water and carbon dioxide. Nitrogen, phosphorus, potassium, calcium, magnesium and sulfur are required in large amounts, thus are called macronutrients. Iron, manganese, copper, zinc, boron, chloride, molybdenum are required in relatively small amounts, thus are called micronutrients, or minors.

pH

pH controls the uptake of nutrients. pH is a measure of the concentration of hydrogen (H+) ions, sometimes called protons. The greater the H+ ion concentration, the more acid the solution, hence a lower pH. If the pH is not in the desired range, individual nutrients can not be taken up, creating a nutrient deficiency, or the nutrient can be taken up too readily, resulting in a nutrient toxicity. These nutrient imbalances will occur even when proper amounts of nutrients are applied to the media, if the pH is too high or too low. The figure below demonstrates the availability of nutrients to plants at different media pH.

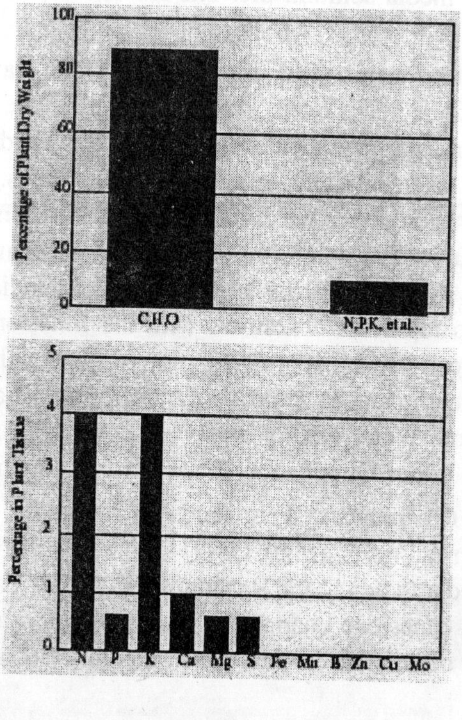

Nitrogen and potassium are readily available at a wide pH range. Although phosphorus is more readily available at a low pH, phosphorus problems are not commonly observed in greenhouse crops. Calcium and magnesium are more readily available at a higher pH. At a low pH, the minor nutrients are readily available. Minor nutrient toxicities are relatively common at a low pH (<5.8), while deficiencies frequently occur at a high pH (>6.5)

Factors Affecting Media Solution pH

Water Quality/Alkalinity: Alkalinity is one measure of the quality of water used for irrigation. Alkalinity is the measure of the concentration of bicarbonates and carbonates in water which determine the water's capacity to neutralize acids. In other words, irrigating with bicarbonates in water is equivalent to applying lime with each irrigation. The bicarbonates react with hydrogen ions and remove them from solution. This process effectively decreases the H+ ion concentration in the media and thus increases the media solution pH. The reverse situation can also occur.

Very pure water (low bicarbonates) can cause media solution pH to decrease over time. The pH drops, because there may not be enough bicarbonates to absorb excess hydrogen ions. Thus, the H^+ concentration in the media increases. The most common solution for pure water sources is to increase the amount of pulverized dolomitic limestone incorporated into the media prior to transplanting plants into the media. Another solution is to top-dress containers with the limestone. Finally, bicarbonate can be added to irrigation water in the form of potassium bicarbonate to improve the buffering capacity of the media solution. Water quality issues are covered in more depth in Irrigation Water Quality for Greenhouse Production.

Media Components: Peat tends to be acidic. Pulverized dolomitic limestone (CaMg(CO3)2) is incorporated into most amended media to adjust the starting pH to ~6.0. Coarser grades of dolomitic limestone change the media pH more slowly, and thus are not often used in peat-based media. A relatively new, but popular media component, coconut coir, is less acidic than peat, so less limestone needs to be used. The role of media in a greenhouse nutritional program is covered in more depth in Growing Media Quality for Greenhouse Production.

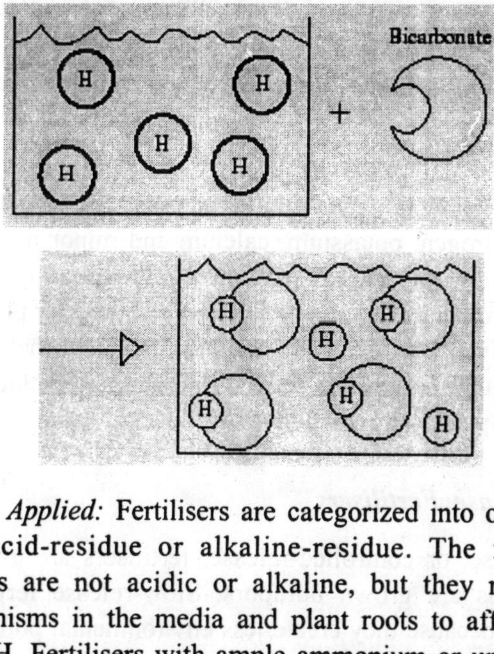

Fertilisers Applied: Fertilisers are categorized into one of two groups: acid-residue or alkaline-residue. The fertilisers themselves are not acidic or alkaline, but they react with microorganisms in the media and plant roots to affect media solution pH. Fertilisers with ample ammonium or urea tend to acidify the media, i.e., lower the pH. Fertilisers with ample nitrates tend to raise the pH of the media solution slowly over time.

FERTILISERS AND FERTILISATION

Water-Soluble Fertilisers

Most greenhouse fertilisation programs rely on water-soluble fertilisers to provide most of the nutrients required for plant growth. Water-soluble fertilisers are often applied at each irrigation. This is referred to as a constant liquid fertiliser (CLF) program. A specific fertiliser program must be developed around the irrigation water, media and crops grown. Following is a typical CLF program:

— 200 ppm N from 20-10-20 Peat-Lite Special applied each irrigation for one week.

— 200 ppm N from 15-0-15 applied each irrigation the following week.

— 100 ppm Mg from Epsom salts (magnesium sulfate) applied once.

Repeat: The 20-10-20 Peat-Lite Special supplies nitrogen, phosphorus, potassium and minor nutrients. The 15-0-15 fertiliser supplies nitrogen, potassium, calcium and minor nutrients. The epsom salts supply magnesium and sulfur. Rotating these three products provides all essential nutrients required for plant growth. Recently available are water-soluble fertilisers that supply all essential nutrients in one fertiliser. Examples include 15-5-15 Cal-Mag Special and 13-2-13 Plug Special.

Slow-Release Fertilisers

Slow-release, or controlled-release, fertilisers are usually used when crops are grown outdoors. Slow-release fertilisers are beneficial because they create less environmental pollution, e.g., fertiliser run-off, when sprinker irrigation is used, and they continue to supply nutrients during rainy weather. Slow-release fertilisers are marketed based on the time of release, for example, 3- to 4-month longevity.

The actual fertiliser release rate is determined by the temperature and water content of the media. Therefore, the actual effective release time of the fertiliser may vary from the labeled time. Slow-release fertilisers can be incorporated into the media prior to filling the containers or top-dressed after planting. Slow-release fertilisers are often incorporated into the media for garden mum production as an insurance policy against rainy weather.

Fertiliser Labels

Nutrient Analysis: The fertiliser analysis indicates the percentage of a particular nutrient contained within the fertiliser. The fertiliser analysis typically refers to the percentage of nitrogen (N), phosphate (P_2O_5) and potash (K_2O) contained in a given fertiliser. A balanced fertiliser should provide nutrients in amounts relative to plant requirements. Since nitrogen and potassium are used in relatively similar amounts, a fertiliser should have a nitrogen to potassium ratio of approximately 1:1. Phosphorus is required to a lesser degree, so the nitrogen-to-phosphorus ratio should be approximately 2:1 to 4:1. Therefore, a 2:1:2 (N-P_2O_5-K_2O) is suitable for most greenhouse crops. An example of this type of fertiliser is 20-10-20. While 20-20-20 is still commonly used, 20-10-20 is preferred, since the N-P_2O_5-K_2O ratio is closer to that required by plants. The extra phosphorus provided by 20-20-20 is usually wasted, thus creating potential environmental concerns.

Nitrogen Form: Nitrogen is provided in three different forms: nitrate-nitrogen (NO_3), ammoniacal-nitrogen (NH_4) and urea-nitrogen. The nitrogen form affects plant growth and media solution pH. Ammoniacal nitrogen, sometimes called ammonium, tends to contribute to "lush" plant growth, for example, greater leaf expansion and stem elongation, whereas nitrate nitrogen produces a "hard" or well-toned and compact plant. High ammonium can be toxic to plants during cold, cloudy growing conditions. Therefore, ammonium and urea should make up <40 percent of the nitrogen during winter months. "Dark-Weather" or "Finisher" fertilisers tend to have high nitrate and low ammonium nitrogen. The following equation demonstrates how to calculate the percentage of the total nitrogen that is in the ammoniacal form. When calculating the percentage nitrate versus ammonium, assume urea will break down to ammonium.

$$\% \text{ N in ammonium form} = (\% \text{ Ammonium} + \% \text{ Urea}) \div \% \text{ Total N} \times 100$$

For example, a 15-5-15 label indicates the following nitrogen breakdown:

Total Nitrogen (N)......15%, 1.20 % Ammoniacal Nitrogen, 11.75 % Nitrate Nitrogen, 2.05 % Urea Nitrogen,

(2.05% Urea + 1.20% Ammonium) ÷ 15% Total N x 100 =
21.7% N in ammonium form

Potential Acidity or Basicity: The potential acidity or basicity indicates how the fertiliser will affect media solution pH. A fertiliser label will indicate that the fertiliser has either a potential acidity or a potential basicity. The potential acidity refers to the fertiliser's tendency to cause the media pH to decrease, while the potential basicity refers to the fertiliser's tendency to cause a media pH increase. The larger the number, the greater the tendency for the media pH to be affected by the fertiliser.

Fertilisers high in ammonium cause the pH to decrease, while fertilisers high in nitrate cause the pH to increase. Fertilisers with a considerable percentage of the nitrogen in the ammonium form tend to leave an acid residue in the media, indicated by the potential acidity. Fertilisers that have low ammonium, and thus high nitrate form of nitrogen, tend to leave an alkaline, residue indicated by the potential basicity. The high-ammonium fertilisers tend to have very little or no calcium or magnesium, while the low-ammonium, alkaline-residue fertilisers contain higher levels of calcium or magnesium.

Proper Dilution Rate: The proper dilution rate is indicated on the fertiliser label and can be tested with a soluble salts meter. The soluble salts concentration of the fertiliser solution increases as the amount of fertiliser increases. For example, 20-10-20 Peat Lite Special will have an electrical conductivity (EC) of 0.33 mmhos/cm for every 50 ppm of nitrogen. Therefore, a fertiliser mixed to provide 250 ppm nitrogen will have an EC of 1.65 mmhos/cm. Each fertiliser has a different soluble salts to nitrogen relationship, so the specific fertiliser label must be examined.

Fertiliser Proportioners

Injectors mix precise volumes of concentrated fertiliser solution and water together. Injectors or proportioners are commonly available in a mixing range of 1:16 to 1:200. For example, 1:100 injection ratio indicates that one gallon of concentrated fertiliser will produce 100 gallons of final fertiliser solution. Injectors allow growers to have a smaller stock tank and mix their fertiliser stock solutions less frequently.

However, not all fertilisers can be mixed together. Calcium and magnesium fertilisers typically can not be mixed with phosphate and sulfate fertilisers while concentrated. A solid precipitate will form in the bottom of the stock tank if the fertilisers are not compatible. Once the individual fertilisers are diluted to their final concentration, then all fertilisers are compatible and thus can be mixed together.

Multiple Injectors

Multiple injectors or multiple-headed injectors can be used to inject incompatible stock solutions. If separate injectors are plumbed serially, i.e., one after the other, then fertiliser stock solutions can be mixed at the same concentration as if one injector is being used. For example, one head can inject calcium nitrate, while the other head injects magnesium sulfate.

However, if two injector heads are placed into one stock solution, then the final concentration delivered to the plants will be twice the desired concentration, unless proper dilution occurs, e.g., mix the stock solution for 100 ppm N if 200 ppm is desired.

Injector accuracy and calibration

It is very common for injectors to lose calibration accuracy over time. Growers should test the calibration accuracy with a soluble salts meter every time a new batch of fertiliser is mixed. If a soluble salts meter is not available, then Calibration Method #2 can be performed.

Calibration Method 1: Use a soluble salts/electrical conductivity (EC) meter to determine concentration of fertiliser coming from the end of the hose. The EC of the water must be subtracted from the EC of fertiliser solution. The correct EC for a given concentration is usually found on the fertiliser label.

Calibration Method 2: Place the siphon into a known volume of solution, e.g., a quart or a gallon. Turn the water on through the injector and fill up a large container, such as a 20- or 40-gallon garbage can. When the siphon has removed all of the solution from the small container, turn off the water. If using a Hozon Injector (1:16), then one gallon of stock solution should empty into 16 gallons final solution. If a 1:100 injector is used, then one quart of stock solution should fill 25 gallons of final solution. Table 2 provides assistance in calculating the amount of fertiliser to mix to make different stock solutions using different injection ratios.

Table 2: A chart to determine the number of ounces of fertiliser

	Ounces of fertiliser to make 100 ppm nitrogen at an injector ratio of:			
% N	1:16	1:50	1:100	1:200
10	2:16	6.75	13.5	27.0
15	1:44	4.5	9.0	18.0
20	1:08	3.38	6.75	13.5
25	0.86	2.7	5.4	10.8
30	0.72	2.25	4.5	9.0

To use this chart: Select your injector's ratio setting across the top of the columns.

— Select percentage of nitrogen (N) formula used in left column.

— Read across and down to find ounce required per gallon of concentrate.

— Multiply this amount by the number of gallons of concentrate used in your fertiliser stock tank.

The table is based on 100 ppm N. For 150 ppm, multiply amounts to be used by 1.5; for 200 ppm, multiply amounts to be used by 2, etc.

A FERTILISATION PROGRAM

Nutrients can be placed into the media prior to planting, i.e., a pre-plant nutrition program, and during plant growth, i.e., a post-plant nutrition program. Do not forget that irrigation water can also be a significant source of plant nutrients, especially calcium and magnesium. Despite considerable gardening advice to the contrary, specific nutrients do not pro- mote rooting or flowering! Specifically, phosphorus does not promote rooting and potassium does not promote flowering. Excess nitrogen can potentially reduce flowering and produce excessive vegetative growth.

Pre-Plant Nutrition Programs

Nutrients can be supplied in limited quantities, while the media components are being mixed. Calcium and magnesium are provided when dolomitic limestone is used to adjust the starting pH. Phosphorus and sulfur are provided with superphosphate plus gypsum.

Iron, manganese, zinc, copper, boron and molybdenum are provided with micronutrient formulations. Nitrogen and potassium are provided with potassium nitrate. Typical pre-plant recipe for 1 cubic yard of soilless media:

Dolomitic limestone	10 lbs.
Treble Superphosphate	2.25 lbs.
Gypsum	1.5 lbs.
Micromax	1.25 lbs.
Potassium Nitrate	1 lb.

(The additional nitrogen and potassium will last –1 to 2 weeks.)

Post-Plant Nutrition Programs

Most small to medium-sized commercial greenhouses use commercially blended fertilisers for convenience and dependability; however, for some growers it is economical to buy individual fertilisers and mix them together. Some important notes about each of the essential plant nutrients:

Nitrogen (N): Sources of nitrogen are ammonium nitrate, urea, calcium nitrate, potassium nitrate, magnesium nitrate. The concentration applied is determined by the amount of leaching. For example, in a constant liquid feed program using a subirrigation system 100 ppm N may produce adequate growth, while 300 ppm N may be needed if overhead irrigation results in 25 percent leaching.

Crop requirements:

Bedding Plants	Light
New Guinea Impatiens	Light
Geranium	Moderate
Poinsettias	Moderate to heavy
Chrysanthemums	Heavy

Phosphorus (P): Sources are ammonium phosphate, urea phosphate A nitrogen-to-phosphate (P2O5) ratio of 2:1 is acceptable for most crops. Fertilisers with high concentrations of ammonium phosphate, such as 9-45-15, appear to promote stem stretching.

Potassium (K): Sources are potassium nitrate, potassium sulfate A nitrogen-to-potash (K2O) ratio of 1:1 is acceptable for most crops.

Calcium (Ca) and Magnesium (Mg): Sources are dolomitic limestone, irrigation water, calcium nitrate, magnesium sulfate, magnesium nitrate. Calcium and magnesium provided by dolomitic limestone are released slowly over several months. These two nutrients can have an antagonistic relationship, thus a Ca:Mg ratio of 3:1 to 5:1 is desirable. Calcium and magnesium are commonly found in irrigation water, especially high alkalinity water. Calcium and magnesium deficiencies are most common when the pH is low. Calcium and magnesium fertilisers can not be mixed in the concentrated form with phosphate or sulfate fertilisers, thus calcium and magnesium are frequently omitted from commercial fertilisers.

A few relatively new fertilisers contain calcium and magnesium along with the nitrogen, phosphorus and potassium. These fertilisers often list five numbers in the analysis. These numbers represent N, P2O5, K2O, Ca and Mg, respectively. For example, (15-5-15-5-2, 14-0-14-6-3, 13-2-13-6-3). Epsom salts are 10 percent magnesium by weight.

Micronutrients: Micronutrients are sold in different formulations; for example, Micromax, Esmigran and Soluble Trace Element Mix contain only inorganic sources, while Compound 111 contains chelated sources. Chelated forms are superior in that the micronutrients are more soluble, therefore more readily available to the plant. Consequently, chelated micronutrients are applied at lower rates. Compound 111 and STEM are labeled for use in constant liquid feed programs. The rates are based on adding a certain amount of micronutrient mix per 100 ppm of N used in the fertilisation program. Micronutrient deficiencies are closely related to media pH. High pH (greater than 6.5) can produce deficiencies, while low pH (less than 5.8) can cause toxicities. Adjusting the media pH is the best solution to avoid micronutrient toxicities or deficiencies.

CHAPTER 8

TOOLS IN ORNAMENTAL PLANT BREEDING

Interspecific hybridisation is the most important source for variation in ornamental breeding. Interspecific hybrids have the potential to capture hybrid vigour as well as combine traits that do not occur within a single species. Because breeder rather than geneticist always want to add a new type of characteristics to the current cultivars, interspecific hybridisation is indispensable to combine diverse gene pools. Thus interspecific or intergeneric hybrids have the enormous potential to extend not only their qualitative but also quantitative traits such as the type of flower, plant phenotypes, other single dominant traits from parent species with an environmental adaptation or improve the flower vase life. While natural hybrids can exist between species whose flowering times overlap, pre- and post- fertilisation barriers hinder the frequency of these hybrids.

In lily interspecific hybridisation has greatly contributed to the breeding for the commercial Longiflorum-Asiatic hybrids, which are becoming more important on the market. The point of current breeding of important horticultural hybrids has two distinct points. One is interspecific hybrid and the other is polyploids origin. In fact, more than 60-70 % are polyploid origin in angiosperm. However, the systematic use of this genetic phenomenon is more important to commercial breeding. As in natural condition the polyploid gametes have originated spontaneously from interspecific hybrids. In those early days the knowledge regarding on the polyploidisation was very poor, the horticultural breeders were successful in creating considerable genetic variation. These efforts were subsequently responsible for selecting thousands of horticultural varieties in some of the crops.

However, the genetic consequence of the interspecific hybridisation has not been dealt in finding the efficient way for the introgression. Therefore, it is important to elucidate the cytogenetic composition of many of these crops in order to practice a more systematic breeding. The number of produced interspecific hybrids and their backcross progenies solely depended on the seed set after normal sexual hybridisation until plant tissue culture techniques has not been settled. However, the prevalence of pre- and post-fertilisation barriers to crossing species, as well as the high degree of sterility of the F1-hybrids, hinder introgression between two species and, when successful, it was a highly laborious and frustrating process.

Nevertheless, the introduction of ovule culture and embryo-rescue methods has greatly facilitated wide hybridisation followed by backcrossing in a wide variety of plant species. Although these techniques opened the possibilities for the production of interspecific hybrids and backcross progenies on a large scale, the next important steps of selecting genotypes from the backcross derivatives, i.e., selection methods, were required. The development of polyploid forms of breeding in ornamental plants has also been important.

In Delphinium, new cultivars, which resulted from hybridisation among plants with different ploidy levels, have appeared in recent years. A combination of in situ hybridisation and molecular markers can be not only useful for monitoring the hybrids and back cross progenies for introgression but also for elucidating the modes of origin of 2n-gametes, the extent of genetic recombination and the phylogeny of the species and hybrid cultivars.

INTERSPECIFIC HYBRIDISATION

Pre-Fertilisation Barriers

In order to introduce valuable economic traits such as disease resistance, flower shape and colour, from wild species into the cultivar assortment, it is essential to overcome interspecific crossing barriers. Several techniques hitherto, as the cut style method, the grafted-style method and the in vitro isolated ovule pollination technique, have been developed to overcome pre-fertilisation barriers. To overcome prefertilisation barrier, the environmental conditions such as high temperature, hormone treatment, specific protein and exudates and lipids can be applied.

Post-Fertilisation Barriers

Post-fertilisation barriers can be circumvented by in vitro rescue methods as embryo, ovary-slice and ovule culture in case of lily and other important ornamentals. Interspecific hybridisation is highly relied on the genotype combination among species. Fertilisation can be classified into embryo and endosperm formation. Failure of the fertilisation in interspecific hybridisation is assumed mainly due to genetic discrepancy and chromosome degeneration during cell division of the zygote. In interspecific hybridisation of Lilium, the embryo and endosperm formation and development can be categorized in different groups. Because of the development of embryo is highly affected by the development of endosperm. If there is no endosperm formation, normally the embryo cannot survive anymore.

In this case, very early embryo rescue such as ovary slice culture may helpful. There are several obligations in interspecific hybridisation, which are important to know why interspecific hybridisation cannot be successful. Firstly, fertilisation has to be successful both embryo and endosperm. Secondly, embryo development has to be checked.

Thirdly, endosperm development has to be monitored. Based on the results of certain cross combination, suitable rescue method can be applied. In this case, the time of embryo rescue looks important but it is not always effective for the embryo rescue. For unknown reasons embryos degenerate in a very early stage even several days after fertilisationalthough it is not clear whether or not the fertilisation has been occurred normally. In the case of early embryo degeneration the ovary slice culture method is preferred. Many cases showed that endosperm grows longer than the embryo. When the embryos are degenerated in a middle stage of the development, ovule culture can be applied.

In all crosses made in Delphinium by Honda & Tsutsui, pollen germination was normal on the stigma, and numerous pollen tubes reached the upper part of the ovary. The observations may confirm that although fertilisation itself might occur normally, a post-fertilisation barrier arose. Hybrid embryos from these crosses probably were aborted due to the lack of endosperm development. Interspecific hybrids in ornamental crops have been reported in Lilium, Lupinus, Sandersonia, Alstroemeria, Tulipa and many others crops.

Integrated techniques have been considered to handle both pre- and post-fertilisation barriers. Unlike the other techniques, which retain the zone of inhibition (stigma and style) and manipulate pollen germination and pollen tube growth to overcome pre-fertilisation barriers, in vitro pollination brings pollen grains in direct contact with the ovules, and is, therefore, considered more effective. In Lilium various combinations of in vitro pollination (cut-style and grafted-style method) and embryo rescue (ovary, ovule and embryo culture, placental pollination), were applied in order to control the whole fertilisation process. This resulted into a range of new interspecific hybrids.

Using these techniques wide interspecific lily crosses with species and cultivars from the different sections of the genus

Lilium (L. longiflorum, L. henryi, L. canadense L. concolor, L. dauricum, L. candidum, L. rubellum, L. martagon, Asiatic and Oriental hybrids) could be made. Hybrids originating from intersectional crosses (e.g. L. longiflorum x L. concolor, L. longiflorum x L. dauricum, L. longiflorum x L. henryi, L. longiflorum x L. martagon, L. longiflorum x L. candidum, L. longiflorum x Asiatic hybrids (LA), L. longiflorum x Oriental hybrids (LO), L. longiflorum x L. rubellum, L. longiflorum x L. canadense, Oriental x L. auratum x L. henryi, L. longiflorum x (L. longiflorum x L. rubellum) Oriental x Asiatic hybrids (OA) and L. henryi x L. candidum) have been produced. Especially the Oriental x Asiatic hybrids are a break-through in lily breeding and a promise for the future. Similar results were obtained from interspecific crosses in Tulipa and intergeneric crosses between Nerine and Amaryllis. Until now in vitro fertilisation for bulbous crops has not been achieved using isolated sperms and eggs.

POLYPLOIDISATION

The introgression of summer-hardiness traits into Impatiense hawkeri germplasm has been hampered by sterility in the interspecific hybrids and in their offspring. Amphidiploidy has been used to create male and female fertility, but this has made introgression more difficult because segregation in progenies is hindered. Naturally occurring pollen fertility has been found sporadically among interspecific hybrids without the use of amphidiploidy. Methods for estimating pollen fertility include counting unstained, normal-looking pollen grains or grains stained with acetocarmine.

Acetocarmine or Aceto orcein and FDA (Fluoro diacetate) have been used for the staining of fertile pollen in many crops. Especially pollen from interspecific or intergeneric hybrids can be checked for their fertility. Under the stereo microscope, pollen from Liliaceae can be observed easily for their size as n- and/ or 2n-pollen. Because pollen of distantly related interspecific or intergeneric hybrids can not be germinated in a normal condition, however, a low percentage pollen are able to germinate. In this case we have a clue that these viable pollen is not reduced but exists of only unreduced pollen. Most of unreduced gametes are produced by FDR with maximum heterozygosity. Theoretical point of view of FDR, there are some extent of homoeologous recombination between parental genomes.

Mitotic Polyploidisation

In almost all cases, the F_1 hybrids of distantly related species are highly sterile. In most cases sterility is resulted from the disturbed chromosome division during meiosis which leads to the formation of spores with unbalanced chromosome constitution leading to sterility. The most widely used method of restoring fertility in interspecific hybrids is by doubling the chromosome number of the F_1 in order to produce an amphidiploid. Because of homologous chromosomes pair with each other, meiosis will be normal. The pollen can be viable with balanced euploid number.

One great drawback of this approach is, however, that because of the preferential pairing of chromosomes between the constituent genomes of the hybrid, the possibility for homoeologous chromosome pairing and crossing- over is dramatically reduced. Since homoeologous recombination is a crucial prerequisite for introgression of specific desirable characters into a cultivar, chromosome doubling of the F_1 hybrid is not a desirable method. However, whole chromosomes can be added to the backcross progenies as has been demonstrated in the case of diploid (2n=2x=24) hybrid of Lilium longiflorum × L. rubellum and their BC_1 progeny using GISH. However, in this intersectional hybrid there is evidence for homoeologous chromosome pairing and crossing-over but such recombinant products cannot be transmitted into next generation due to sterility.

Although there is a drawback being reduced homoeologous recombination, there are many chromosomes from donor into recipient plant. In this case whole chromosomes are introgressed rather than chromosome segments. As will be described below for a different hybrid, through the use of 2n-gametes homoeologous recombinant products can be recovered in the backcross progeny.

Meiotic Polyploidisation

Occasionally, plant species and interspecific hybrids produce 2n-gametes instead of the normally expected haploid, or n-gametes. When they occur in interspecific hybrids in some extent of viability, 2n-gametes can be used for the production of sexual progeny either through crossing or selfing. Progenies in these cases consist of polyploids and offer an alternative to somatic chromosome doubling. If 2n-gametes from only one parent was crossed to diploid counterpart parent, 3x progeny will be produced in general. Indeed, meiotic doubling has been practised successfully in some of the crops such as Alstroemeria and Lilium.

Unlike mitotic doubling, sexual polyploidisation saves time that would have required for colchicine treatment. But the most important advantage of meiotic doubling is that homoeologous recombination occurs during meiosis in the diploid interspecific hybrid and many alien chromosome segments can be transmitted. Because, unlike in a somatically doubled allotetraploid, the homoeologous chromosomes are forced to pair during meiosis diploid interspecific hybrid. This forms the basis for the occurrence of intergenomic recombination in the 2n-gametes. Such homoeologous recombination due to crossing-over has been clearly demonstrated in the case of Alstroemeria, Lilium and Gasteria lutzii × Aloe aristat. There are two main mechanisms known so far viz. FDR and SDR.

Recently a third mechanism 'IMR' was reported by Lim et al. in lily. Because homoeologous chromosome pairing can be highly variable in interspecific hybrids, ranging from the formation of only univalents, i.e., without any crossing-over, to complete pairing as bivalents (high degree of crossing-over) in some of the spore mother cell, several mechanisms can be considered. The most common mechanism of 2n-gamete formation is FDR. There are two types of FDR gametes, so called, FDR with homoeologous recombination and FDR without homoeologous recombination. It is interesting that FDR with homoeologous recombination is more often than FDR without homoeologous recombination analysed in lily BC_1 plants.

In view of the occurrence of different types of meiotic restitution mechanisms in distant hybrids, it is possible to predict the potential outcome of using 2n-gametes in breeding provided we have knowledge of the modes of their origin. From the literature on the occurrence of 2n-gametes in plants it is clear that they occur in almost all plants but their frequencies vary greatly. There are claims that different mechanisms of nuclear restitution are determined by single genes in crops like potato and many other crops.

When both n- and 2n-gametes are produced in a fertile diploid plant, the progeny consists of a mixture of diploids and polyploids. The number of diploids is normally higher than that of triploids because of counterpart gamete is also haploid so that embryo-endosperm ratio will be stable. The selection of polyploids will be laborious in this case. However, in a considerable number of plant species a phenomenon called 'triploid block' operates and in these the selection of polyploids can be highly efficient. As regards the second question on how to proceed with triploids and allotetraploids derived from uni- and bi-lateral sexual polyploidisation, there is considerable literature on using both triploids and allotetraploids but certain aspects have been ignored in the past.

Generally triploids are regarded as sterile and of little interest in breeding. However in vegetative propagated ornamentals like Alstoemeria and lily the triploid ploidy level is preferred over diploids and tetraploids. A cursory survey indicates that triploids have been used successfully as parents in planned crossings as well as during spontaneous polyploidisation, e.g., Narcissus, extensively. An interesting feature of triploids is that it is possible to generate progenies with near diploid (circa diploid) or tetraploid chromosome constitution by making 3x-2x (or reciprocal) and 3x-4x (or reciprocal) respectively.

The progeny with a near diploid constitution are potentially useful for breeding at the diploid level as has been demonstrated in the case of analytic breeding in potato and other polysomic polyploids. Like in other plant species, near diploid progenies have been obtained by crossing a triploid Lilium hybrid (2n=3x=36) with diploid (2n=2x=24) genotypes. The different ploidy levels observed between 3x-2x and 3x-4x crosses are possibly caused by preferential fertilisation or survival resulting in a different ratio of chromosome numbers between the embryo and endosperm.

Though Lilium has tetrasporic, eight nucleate type of embryo sac formation (Fritillaria type), the observed difference between the progeny types in 3x-2x and 3x-4x crosses is comparable to that of observed in monosporic eight nucleate type (Polygonum type) that predominate in most genera of Angiosperm. An important feature of the genome constitution of the progenies was that the homoeologous recombinant chromosomes were transmitted intact from BC_1 to BC_2 progenies in variable numbers. In addition, there was evidence for the occurrence of new homoeologous recombinations in the triploid BC_1.

One putative hybrid (2n=24) from Delphinium hybridum 'Galahad' (2n=32) × D. grandiflorum (2n=16) showed that some of phenotypic characters were similar to D. hybridum 'Galahad', others are look like D. grandiflorum. The leaf shape of this hybrid was intermediate between those of the parents. The percentage of viable pollen of parents was about 76 %, whereas that of the D. hybridum 'Galahad' × D. grandiflorum hybrid was only 8.5 %. The hybrid failed to yield viable seeds either by self-pollination and back crossing.

Whereas wild allopolyploid plants are well adapted for a long time, man-made allopolyploids are typically unstable. Some progenies of 3x × 2x cross of interspecific lily hybrids died in vitro stage, while others grew very vividly. This instability refers to the appearance of unexpected changes in phenotype, such as homeotic transformation and lethality, and in genome structure, such as chromosomal rearrangements and changes in the number and distribution of repeated DNA sequences within heterochromatin. However, the molecular mechanisms underlying these adaptive steps remain obscure.

Interspecific hybrids between distant relatives are usually sterile. Allopolyploids represent a special type of hybrid that contain two parental genomes undergo little intergenomic recombination and thus maintain their integrity through sexual generations. Domestication of closely related interspecific hybrid might be more reliable. Because the homoeologous chromosomes between parental species or subspecies are able to pair at higher level so that many breakpoints of homoeologous recombination are possible just nearly the same act as homologous intra-specific recombination. That's how natural allopolyploids have been adapted more stably rather than man-made allopolyploids.

In an ancient era some species in the same genus were more homologous in their genomes, that may affect more chance being natural allopolyploids. Therefore, karyotypic stability is achieved at the expense of the evolutionary flexibility provided by unhindered recombination between the parental genomes. One factor that favours allopolyploids may be heterosis generated by the combination of homoeologous genes. Although the prominent role of polyploids in plant breeding has long been recognized, underlying mechanisms of gene expression, causing the novel variation are still poorly understood.

FLORAL LARCENY AND PLANT REPRODUCTION

Insects and other animals pollinate 90% of some 300,000 species of angiosperms worldwide, an end-product of >100 million years of evolution and adaptive radiation by both the animals and plants. But this evolutionary history began with interactions in which the animals were opportunistic or obligatory herbivores, and modern-day mutualisms between plants and pollinators are not inherently cooperative.

Instead, they represent compromises between the requirements of plants for sexual reproduction versus those of pollinators for food and other floral resources. These compromises can be exploited. Nowhere is this clearer than in the phenomenon of floral larceny, an umbrella term covering a range of behaviors by which animals obtain rewards from flowers in ways that appear to effect no pollination and/or to damage the flowers.

Inouye in broadly classifying larcenists as "behavioral robbers" (hereafter "robbers"), who pierce flowers to extract nectar rather than legitimately entering them, and "thieves", who enter flowers as pollinators do but transfer little or no pollen, usually due to a mismatch with the flower's morphology. Recognition that flower visitors include facultative and obligatory larcenists dates at least to Herman Müller and Charles Darwin, who discussed robbing behavior in detail over a century ago. Since then, evidence for the ubiquity of robbing and thievery has accumulated. However, a clear picture of its importance for individual fitness and for population- and communitylevel dynamics is just beginning to emerge.

The net effects of cheating for female plant reproduction can be negative, positive, or neutral. The direct negative effects of nectar robbing, via damage to reproductive organs and/or antagonistic interactions in which robbers drive away legitimate pollinators, are easy to envision. Conversely, some nectar robbers actually have a direct positive effect by pollinating at least a fraction of the flowers they rob. Pollination in such cases ranges from occasional and accidental at one extreme, to regular and "intentional".

To understand this range of behaviors of flower visitors, all the behaviors occur because they are efficient ways to obtain resources. This leads to the prediction that larceny will be obligatory for an animal in some cases and, in other cases, facultative, depending on the rewards provided in the flowers visited and on the rest of the floral environment (and hence other resources) available to that animal. Furthermore, it may be efficient for an animal to behave as a robber or thief with respect to one floral resource and behave in a "legitimate" fashion with respect to another.

For example, bumblebees systematically rob older flowers of Mertensia paniculata which contain nectar, but buzz-pollinate young flowers which contain only pollen. Or, the intentional collection of one resource may engender pollination. For example, female carpenter bees pierce flowers of ocotillo to obtain nectar, but pollinate while collecting pollen from the same flowers. In short, close inspection can reveal combinations of behaviors and net fitness effects that are unexpected if one presumes that animals act only in stereotypic ways that can be placed into a few convenient categories.

Additionally, robbers and thieves may have more subtle indirect effects on plant fitness, for example by altering the interaction between plants and pollinators. The result may be to decrease plant fitness if larceny reduces floral rewards sufficiently for pollinators to avoid a plant or to desert it after a brief visit. Or the result may be to increase fitness if each pollinator visits fewer flowers per plant, so long as there is a sufficiently high fitness cost of within-plant selfing. Maloof and Inouye reviewed nectar robbing (but not thievery) and concluded that robbers are about equally likely to increase or decrease plant fitness.

However, their analysis did not take into account the magnitude of fitness effects and, therefore, suffers from a form of "vote counting" in which a judgment of positive and negative effects is based only on the statistical significance of outcomes. An effect size (d) that expresses the difference in mean female reproductive success in plants or flowers that did versus did not experience larceny, divided by their pooled SD. A positive effect size indicates positive effects of robbing and vice versa. An absolute value of 0.2 is considered a small effect, 0.5 medium, 0.8 large, and ≥1.0 very large.

The cumulative effect size across studies to determine the overall impact of larcenists and then compared effect sizes across studies of different types of larcenist-pollinator combinations (bird-bird, insect-bird, and insect-insect), different types of larcenists (robber versus thief), different research approaches (experimental versus natural larceny), and different interactive outcomes (positive versus negative effects of robbing). These comparisons were analyzed statistically using a between-class homogeneity statistic (Q_b) which has a χ^2 distribution with df equal to the number of classes minus one.

To compare studies that yielded positive versus negative effect sizes, here used the absolute value of effect size to compare the magnitude of effect size rather than the direction. Overall, floral larceny had a weak but negative effect on female reproductive success (d=-0.27). Individual values of d ranged from -3.01 to +2.74 (Fig. 1). On average, studies finding negative effects of floral larceny reported stronger fitness consequences than those finding positive effects (Fig. 1). The type of larcenist-pollinator combination had a significant impact (Q_b=17.25, *df*=2, P<0.0001).

In plants that were bird-pollinated, insect larcenists had a weak negative effect (*d*=–0.25) on female reproduction, while avian larcenists had a relatively strong negative effect (*d*=–1.07). For insect-pollinated plants, floral larceny by other insects had a weak positive effect (d=+0.23), apparently driven by cases in which robbers pollinate flowers from which they steal nectar, either accidentally or "intentionally". No difference in the effect of nectar robbers versus nectar thieves (Q_b=2.01, df=1, P>0.05). Finally, significantly larger effect sizes in studies that manipulated floral larceny experimentally than in those that were purely observational (d=0.87 versus 0.25; Q_b=32.06, df=1, P<0.0001).

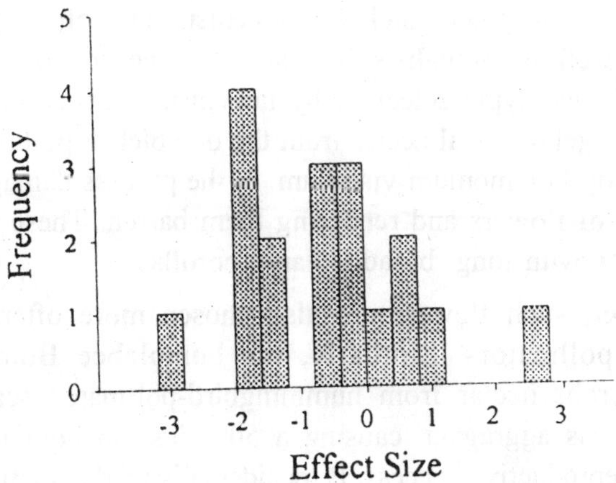

Figure 1: Frequency distribution of effect sizes from a meta-analysis on the effect of floral larceny on female plant reproduction. Negative values indicate negative effects of larceny on reproduction, and vice versa

For example, pollinators might not discriminate between robbed and unrobbed flowers on the same plant, but might discriminate between plants with high versus low robbing. Comparing robbed and unrobbed flowers on plants that have variable levels of robbing may render per-flower effects uninterpretable.

Floral larceny overall has a weak negative effect on female reproductive success of plants. The effect seems dependent on the types of robbers, thieves, and pollinators that interact, and on the reproductive biology of the plants. Manipulate floral larceny to explore the subtleties of direct and indirect interactions, averaged across temporal and spatial variation. Only this kind of effort will allow us to establish clear trends.

NATURAL SELECTION

The evolution of floral traits surely must be understood, in part, with reference to pollinators. But from the plant's perspective, the selection environment includes larcenists, and other plant enemies, as well as mutualists. To date we have precious few estimates of phenotypic selection by larcenists. Ants, Formica neorufibarbus gelida, steal nectar from the bumblebee-pollinated alpine skypilot, Polemonium viscosum, in the process damaging female parts of flowers and rendering them barren. These ants prefer flowers with long, broadly flared corollas.

However, such flowers are also chosen more often by bumblebee pollinators. Similarly, the bumblebee Bombus occidentalis robs nectar from hummingbird-pollinated scarlet gilia, Ipomopsis aggregata, causing a 50% loss in both male and female reproductive success. B. occidentalis workers emerge in midsummer, so plants that bloom later in the season suffer higher levels of robbing than those that bloom early. At the same time, however, later flowering individuals of I. aggregata may receive higher levels of pollination because they avoid competition with an earlier flowering species for the services of hummingbird pollinators.

In both of these examples, plants face tradeoffs in attracting pollinators while avoiding nectar robbers. In the case of skypilot, robbing ants and pollinating bumblebees both prefer plants with broadly flared flowers and, therefore, select in opposite directions on corolla flare. In the case of scarlet gilia, plants that bloom mid-season experience high levels both of robbing and of pollination; here robbers and pollinators may select in opposite directions on flowering phenology.

The net direction and magnitude of selection will likely depend on the relative abundances in space and time of robbers and pollinators. The possibility that natural selection from larcenists can oppose selection from pollinators. If so, we may have one explanation for the observation that substantial variation in floral phenotype is maintained in natural populations.

PLANT POPULATION

For cheaters to affect plant populations, they must influence whole-plant seed production, and there must be a direct link between seed production and subsequent population size. These links are only now beginning to be explored. Increases or decreases in seed production will translate into subsequent numbers of reproducing offspring one generation later. This assumption may be reasonable, but it is not automatically so since other factors, such as limited safe sites for seedling germination, can override seed numbers in determining population size.

Plant enemies other than robbers have population-level impacts. The spatial pattern of seed dispersal by birds accurately predicted the subsequent location of seedlings of a Mediterranean tree in a shrubland site, but not in a forest site, where postdispersal seed predation played a larger role. Individuals of Platte thistle (Cirsium canescens) that were protected experimentally from inflorescence-feeding insects produced more seeds, seedlings, and flowering adults than those not protected.

In figure 2, (a), and mean seed production per plant (+SE) (b), in two sites with and three sites without Linaria vulgaris, a sympatric species that shares a common nectar-robbing bumblebee. Pooled across sites, the presence of L. vulgaris significantly reduced robbing of I. aggregata (ANOVA on arcsine square-root transformed values: $F_{1,159}$=19.1, P<0.0001) and significantly increased seed production of this species (ANOVA on natural log-transformed values: $F_{1,159}$=40.1, P<0.0001)

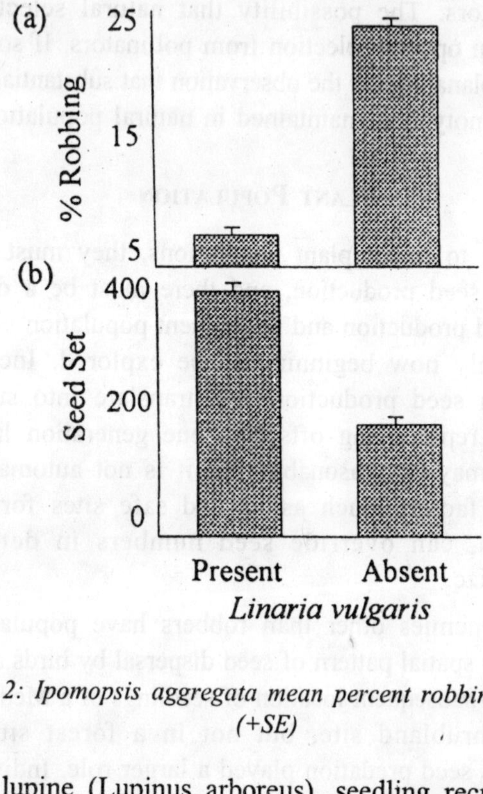

Figure 2: Ipomopsis aggregata mean percent robbing per plant (+SE)

In bush lupine (Lupinus arboreus), seedling recruitment was limited in some areas by the removal of seeds by granivores. The perennial wildflower Ipomopsis aggregata and its nectarrobbing bumblebee, Bombus occidentalis - for which both nectar robbing and seed input have been experimentally manipulated and subsequent population growth examined. As previously noted, heavy robbing by B. occidentalis results in a 50% decline in seed production by I. aggregata.

In experimental plantings, a 50% reduction in seed input indeed yielded virtually a 50% reduction in numbers of new seedlings and first-year juveniles. Because I. aggregata is a semelparous perennial that flowers in some cases only after 7 years or more. However, they do suggest that we may ultimately be able to show for this species that robbing influences the next generation of plants.

COMMUNITY ECOLOGY

The interactions between a plant species, its pollinators, and its larcenists are embedded in a larger web of interactions. Robbed plants grow sympatrically with other plant species that share larcenists and/or pollinators. If two or more plant species share the same nectar robber, they may facilitate each other's robbing. One variant, akin to pollinator facilitation, is sequential facilitation in which sequentially flowering species facilitate each other's robbing through mutual support of robber populations.

Alternatively, if robbers prefer one species, the less-preferred may benefit - an outcome akin to associational resistance to herbivores. Finally, community members may have no effect on each other if robbing intensity is independent of plant density and robbers show little or no preference for any species.

Robbing rates of I. aggregata were 68% lower in the presence of L. vulgaris than in its absence (Fig. 2A), with L. vulgaris receiving 60% more robbing than I. aggregata where the two species grew together. Hence L. vulgaris appeared to have an indirect positive effect on I. aggregata by luring away robbers, as indicated by higher seed set of I. aggregata when it grew with L. vulgaris (Fig. 2B). However, the long-term outcome of the interactions between I. aggregata and L. vulgaris will also depend on the degree to which each population is seed limited, and on the extent of interspecific competition for space.

Suppose that two plant species compete for pollination services and one loses floral rewards (e.g., nectar) to larcenists. Pollinators may avoid foraging on the less-rewarding species and so increase the reproductive success of the unrobbed species - an indirect positive effect through behavioral modification of shared pollinators. Once again, I. aggregata provides a possible example. Broad-tailed and rufous hummingbirds prefer I. aggregata as a nectar source when it co-occurs with paintbrush, Castilleja linariaefolia, another hummingbird-pollinated plant. However, of the two, B. occidentalis robs only I. aggregata.

In so doing, B. occidentalis can shift the preference of hummingbirds toward C. linariaefolia. Preliminary results suggest that seed set of C. linariaefolia is higher in the presence of robbed I. aggregata than in the presence of unrobbed I. aggregata. These examples illustrate some of the ways in which robbers may influence community-level dynamics. In addition, robbers may affect the growth or stability of pollinator populations and the relative importance of one group of pollinators versus another. The complete suite of community-level effects of nectar robbers is so far unstudied.

STABILITY OF FOOD WEBS

Evidence is accumulating that the interactions between flowers and their animal visitors are web-like in structure, at least in temperate ecosystems. Thus, the flowers of many plant species in a community are visited by a diverse array of insects and other animals, and conversely, many animals visit a diverse array of plant species. Indeed, even morphologically specialized flowers often are visited by a variety of animals. At one extreme, all flower visitors might be pollinators, and flower-visitation webs could perfectly correspond to pollination webs.

But the more likely truth is that some flower visitors are larcenists, even leaving aside any behavioral robbers. Thievery by some flower visitors may be obligatory; these animals simply do not "fit" flowers morphologically and so obtain rewards without transferring pollen. Or thievery may be facultative in the sense that the fitness value of visits to the plants depends on the presence or absence of other potential pollinators, and possibly other factors as yet unexplored.

If so, these animals should not be excluded from further consideration: witness their important roles, as outlined above, in plant reproduction, floral evolution, population dynamics, and community-level interactions. Furthermore, thieves may play a role in the stability of communities they belong to. Various aspects of stability, such as spatial and temporal fluctuation in species' abundances, may depend critically on the richness of interactions between trophic levels (in this case, between primary producers and primary consumers) as well as on the strength and direction of these interactions.

For example, theoretical models predict that weak negative interactions reduce temporal variance but increase spatial variance in species' abundances. Furthermore, models of two trophic levels suggest that negative interactions are more stabilizing than positive interactions. Hence, both high connectance of flower-visitor webs, which will reduce the strength of many individual interactions, and the presence of thieves (i.e., negative interactions) in the webs, may contribute to the long-term persistence of these systems.

Chapter 9

Fertilisation of Woody Plants

Urban vegetation include improved physical and mental health and well-being, aesthetic environments, increased property values, wildlife habitat enhancement and noise abatement. It is well documented that air pollution can injure trees but within certain concentration levels plants can effectively act as a sink for airborne pollutants and reduce air quality problems. The magnitude of this impact can be significant, given that forest ecosystems may be the major sink for pollutants within terrestrial ecosystems.

By transpiring water, blocking winds, shading surfaces and modifying the storage and exchanges of energy among urban surfaces, trees affect local climates and consequently energy use in buildings. Trees can reduce building energy by lowering summertime temperatures, shading buildings during the summer, and blocking winter winds. However, trees can also increase building energy use by having their branches shade buildings during the winter, and can increase or decrease building energy use by blocking summertime breezes.

The potential climate modification and building energy use impacts from trees vary with building characteristics, species attributes, climatic region, planting configuration and locations. Trees also have impacts on the hydrology of an urban environment. As areas become developed, the relative amount of impervious surface increases, i.e. parking lots, roads, and buildings begin to dominate over soil and vegetative land cover. Consequently, soil infiltration is reduced, thereby intensifying the volume and rate of runoff, and increasing pollution loads.

These effects cause flooding and water quality problems. Increasing vegetative cover in cities can potentially mitigate many of the hydrologic impacts of urban development. In the past, wildlife management in cities meant animal control, many urban dwellers now value wildlife in their immediate environment as evidenced by increasing positive attitudes toward urban wildlife. Wildlife in cities may offer greater opportunities for environmental education and non-consumptive recreation than remote locations because of the proximity to large numbers of people.

Understanding the environmental, social and economic benefits of trees in cities helps determine the objectives for creating and maintaining an urban forest. Some of the desirable characteristics we can seek in urban trees are: rapid initial growth and then slow growth after maturation, large shade-providing crowns, aesthetic appeal, and robust health with the ability to withstand the many stresses of the urban environment.

GROWTH OF SOIL

Soils store water and nutrients needed by trees. In order to understand tree nutrition and to help determine a need for fertilisation it is crucial to know both the physical and chemical aspects of the soil type in which trees grow. In the Eastern United States, and particularly in the southeast, most of the soils are classified in the Ultisols soil order. In undisturbed areas these soils support an abundant growth of forest species. Ultisols, however, are highly acidic and are often deficient in one or more of the following elements: phosphorus, potassium, calcium, magnesium, sulfur and various micronutrients.

As a rule they are low in effective cation-exchange capacity, resulting in a high leaching potential. In particularly clayey soils, phosphorus is often immobilized and consequently unavailable to trees. While the physical properties of these soils are good, many Ultisols are prone to erosion. The problems associated with these soils are often exacerbated in urban areas where soils are severely disturbed.

SOIL CHARACTERISTICS AND TREE NUTRITION

The amount of nutrients a soil can hold and the availability of these nutrients is greatly influenced by the physical characteristics of the soil - that is its texture, structure, cation-exchange-capacity and pH.

Soil Texture

Soil texture describes the size of the different mineral particles. It relates primarily to particles smaller than 2 mm (.08 in) in diameter - sand, silt and clay. Soil texture significantly influences soil moisture relationships, soil structure, the amount of nutrients adsorbed by the soil, and in turn, plant growth. Generally, the finer the texture of a soil, the greater its capacity to hold nutrients and water; therefore, to supply an equal amount of fertiliser or nutrient, a sandy soil will need more frequent applications of smaller amounts than will a clay soil. It is important to note that a heavy clay may have more total water but water movement may be restricted by poor soil structure and limited soil porosity.

Soil Structure

Soil structure refers to the arrangement of soil particles into units called peds. The form and size of these peds, or aggregates have a profound effect on the porosity and pore-size distribution of the soil. Tree growth is strongly impacted by soil structure because of its effects on the movement of water, nutrient availability, air (aeration) and root growth.

Aeration refers to the amount of pore space occupied by air and aeration status is the primary factor determining the total rooting depth in most soils. In some urban soil environments rooting depth is also affected by actual mechanical and physical impedances and barriers. Plants require oxygen for root growth, which is obtained from the atmosphere diffusing into the soil under a partial pressure gradient. This diffusion can only occur in air-filled spaces. Hence, in compacted or saturated soils this gaseous diffusion is severely limited and root activity is adversely affected.

Cation Exchange

The soil acts as a reservoir for nutrients in that the surfaces of clay and organic matter particles are negatively charged and are able to adsorb positively charged nutrient cations. The amount of the negatively charged sites in soil is called its cation-exchange-capacity or CEC. It is called this because the cations are loosely held and can be easily absorbed by the roots by exchanging them for hydrogen ions. The hydrogen ions account for the acidity of the soil and offer no direct nutrient value to the plant.

Often lime is added to the soil to neutralize acidity and replace the hydrogen on the soil particles with Ca and Mg if dolomitic limestone is used. The greater the clay content and/or organic matter content, the greater the CEC. The types of clay minerals produced from weathering greatly influences their nutrient storage capability; low for kaolinitic types, high for montmorillonitic types. The amounts of nutrients stored by a soil are directly related to the cation-exchange-capacity, and in turn, the fertility of the soil. Soils with low clay and organic matter such as sands or loamy sands would have low fertility.

pH Range

The hydrogen ion concentration (pH) of the soil solution is a measure of a of a soil's acidity or alkalinity and is highly dependent upon the parent material of the soil, regional precipitation, the organic matter content and the major nutrient elements in the soil. Total moisture available to trees and the length of the growing season play predominant roles in the relation between the pH value of soils and the distribution of plants. The longer the season, the higher is the tolerance of plants toward soil acidity.

Soil pH affects the solubility and availability of soil nutrients; each nutrient's maximum solubility occurs over a certain pH range, outside of which the nutrient is likely to form insoluble compounds. Therefore, soil pH greatly influences nutrient absorption. The pH value of soils is not a constant; it varies with the soil: water ratio and the salt concentration. Lowering the amount of water and increasing the salt concentration both increase the pH value.

White oak (Quercus alba L.), tulip poplar (Liriodendron tulipifera L.) and sycamore (Platanus occidentalis L.), and other hardwoods often grow exceptionally well on fine-textured soils with a pH as low as 4.6, although the recommended optimum ranges for tulip poplar and sycamore are between 6.0 - 7.5. Aslander found that soil nutrient content influenced the distribution of Norway Spruce (Picea abies A. Dietr.), European birch (Betula penduia Roth), oak (Quercus spp.), basswood (Tilia spp.) and beech (Fagus spp.) far more than did the pH value of soils.

Hemlock (Tsuga spp.) and balsam fir (Abies balsamea (L.) Mill.), known to be acid-loving species have also been reported growing successfully on calcareous (calcium or lime containing) soils of pH 7.3 to pH 8.0. If a soil is too acid, adding lime can increase its pH; if a soil is too alkaline, sulfur can be added to decrease its pH. Soil texture, organic matter, material used, and how much the pH is to be changed will all influence the amount of lime or sulfur needed.

When the pH of a mineral soil is 4.5 or below, Al, Fe and Mn are so soluble that they may become toxic to certain plants. Harris believes these elements may become toxic at a pH of as high as 5.5. High acidity has also been shown to inhibit earthworms. Very low pH (3.0 - 5.0) may cause P to be tied up in unavailable iron and aluminum phosphates. This problem is less severe with most trees than with agricultural crops because of the ability of the mycorrhizae associated with trees to obtain P from these compounds.

Soils that are excessively alkaline or have a very high pH (above 7.5) inhibit the solubility of Fe and may lead to chlorosis, particularly of oaks and white pine (Pinus strobus L.), on limestone soils low in organic matter. The same problem may occur with Cu and Zn. Highly alkaline soils of 9.0, as may be the case in many urban soils, result in deficiencies of Mn as well as N. Toxic release of K, S, Ca and Md are also possible at such high pH.

Urban Soils

Under urban conditions, natural soils generally undergo considerable alteration; their physical and chemical characteristics do not resemble their natural conditions. Urban trees are often adversely impacted by removal or disturbance of soil due to construction, which results in the loss of the primary storage site of organic matter and nutrients in the soil profile as well as loss or significant alteration of water storage and movement capability.

Severe disturbance may also result in the inversion of soil materials so that topsoil is buried and nutrients become less available. Infertile subsoil may be left at the surface where the primary root zone would normally develop.

Aeration

Of all soil stresses in urban areas, the most common may be

poor aeration due to soil compaction. Compaction destroys macroporosity; and because pore space is reduced, soil resistance, hardness and bulk density are increased. Bulk density is the weight of a unit volume of soil, including the soil particle volume and the pore volume. Because there is little pore space in compacted soil with high bulk density, root penetration is inhibited, water movement through the soil is slow, and there is usually insufficient oxygen for root growth.

The bulk density in natural surface mineral soil in forests ranges from about 0.5 to 1.3 g/cubic centimeter. Many urban soils have bulk density values ranging from 1.6 - 2.2 g/cubic centimeters. A bulk density of 1.5 - 1.6 is often considered to be the threshold value, above which significant negative growth impacts are likely to occur. It is important to select planting materials that will best tolerate soil compaction for heavily trafficked areas. The rates of species are

— *most severely injured*: sugar maple (Acer saccharum Marsh.), beech (Fagus spp.), flowering dogwood (Cornus florida L.), oak (Quercus spp.) tulip poplar, pines (Pinus spp.) and spruce (Picea spp.).

— *less severely injured*: birch (Betula spp.), hickory (Carya spp.), and hemlock (Tsuga spp.).

— *least injured*: elm (Ulmus spp.), poplar (Populus spp.), willow (Salix spp.), plane tree (Platanus spp.), pin oak (Quercus palustris Muenchh.) and locust (Gleditsia spp., Robinia spp.)

Soil Amendments

Amending the soil with porous, yet durable materials ameliorates compaction. Amendment materials are generally divided into organic and inorganic categories. The organics by their nature tend to be short lived in an aerobic soil system and decompose thus enhancing the cationexchange- capacity and structure of the soil. Several kinds of materials are available as mulches to protect the soil surface in heavily used areas such as coarse wood or bark chips, composted leaves, sawdust, straw, and peat moss. The inorganics tend to be longer lived or even resist decay in the soil system.

When cost is not prohibitive, inorganic mineral materials including marble chips, vermiculite, pumice or gravel may be used. In general, solid mineral materials cushion less than organic mulches but they all contain internal pore space within their matrices that seems to provide additional benefits such as water holding capacity. When using amendments, it is important to consider the percentage by volume required in a soil mixture. The answer will vary depending upon the type of amendment, use of the proposed site, the intensity of use, and the degree of maintenance the site will receive.

Organic matter can absorb up to 400% its own weight in water. Incorporating high amounts (10, 20, 30%) of organic matter by volume into a soil could easily result in an anaerobic soil system. Most often a 5 to a maximum of 10% by volume of organic material is recommended. For inorganic materials, and those areas designed for heavy use, Patterson concludes from his research that the ideal percentage of soil incorporation appears to be between 30% and 40% by volume for areas where use will be heavy and maintenance activities intermittent.

pH of Urban Soils

Urban soils characteristically have near neutral to alkaline pH, which are not optimal for many tree species. The weathering of calciumcontaining mortar, concrete and other building materials mixed or washed into the soil, as well as the application of calcium or sodium chloride as road and sidewalk de-icing compounds in northern latitudes bring about a higher soil pH.

Streeside soils of Syracuse, New York had a pH range of 6.6 to 9.0 with an average of about 8.0. Urban soils of Philadelphia, Pennsylvania ranged from 3.7 to 9.0 with a mean of 7.6. In Berlin, a pH of 8 was observed at streetside and less than 4 within a forest a short distance from the street. Applying granular elemental sulfur is an effective means of lowering pH. Applying organic mulches he contends will also effectively lower pH within a limited depth.

Nutrient Cycling

Nutrient cycling poses another problem in urban soils. In a natural, undisturbed forest situation the nutrient cycling processes are a stable, wellbalanced, self-maintaining system. In urban area the cycle is disturbed in several ways. Often the twig and leaf fall is collected and disposed of in landfills so that a continuous loss of nutrients from the soil occurs. This can also occur from mowing of grass or other understory vegetation. Excessive runoff from compacted soil or paved surfaces may also carry away leaves and soil, depositing them in surface drainage systems.

Input of nutrients to urban soils may occur via wet and dry deposition from the atmosphere on trees or soil; from dogs, cats, birds, etc.; and from mulches or fertiliser in some cases. Both nitrogen and sulfur inputs can cause soil acidification and leaching of basic elements from soil. Low organic matter content disfavors aggregation and prevents good soil structure

Toxic Elements

Urban soils receive varying inputs of toxic materials from automobile exhausts, industrial plant emissions, road-salting, and chemical spraying programs. Increased amounts of lead in road-side soils are well documented. Salt has received widespread attention as a roadside pollutant. A great deal of injury to trees is attributed to sodium chloride excess in the trees, particularly chloride, where salt is used for de-icing in the winter. In addition to direct injury, the uptake of basic elements may be disrupted by excess sodium in the soil. If sodium represents more than 15% of the exchange-capacity, a medium or fine-textured soil will lose its granular structure and become exceedingly impervious to air and water.

Temperature Regimes

It is a well-documented fact that urban development modifies the local climate. Temperatures (often called urban heat island), precipitation, and cloud cover increase relative to surrounding rural environments while wind, humidity, and radiation decrease (Table 1). In many cases the soil is surrounded by large capacity heat-absorbing and re-radiating surfaces, increasing the heat flux to the cooler soil.

Evaporation of water from the soil surface eventually dries it and imposes greater stress upon the plants. Nighttime minimum soil temperatures tend to be high because of the high air temperatures from the heat retention of structures. Plant metabolism rates potentially remain high. Unfortunately, few actual soil temperature data are available for urban soils.

Table 1: Average Changes in Climate Elements Caused by Urbanization.

Element	Comparison with rural environment
Wind Speed	
Annual Mean	20 to 30% less
Extreme Gusts	10 to 20% less
Temperatures	
Annual Mean	0.9 - 1.8 degrees F higher
Winter Average	1.8 - 3.6 degrees F higher
Precipitation	
Total	5 to 10% more
Days with less than 5 mm	10% more
Relative Humidity	
Winter	2% lower
Summer	8% lower

Cloudiness	
Cover	5 to 10% more
Radiation	
Total	15 to 20% less
Sunshine Duration	5 to 15% less
Contaminants	
Condensation Nuclei and Particulates	10 times more
Gaseous Admixtures	5 to 25 times more

Soil temperature is important since it controls the growth environment of roots and soil organisms, and inorganic chemical processes. A warmer temperature increases rates of reaction and biological processes. The rate of organic matter decomposition is increased, provided the necessary organisms are present, and the overall soil-weathering process may be intensified. The latter may have a beneficial effect from the release of nutrients for absorption by roots. Root growth may be extended well into the fall and early winter.

ESSENTIALITY FOR A NUTRIENT

Arnon proposed three criteria of essentiality for a nutrient for higher plants. These criteria are: 1) the organism cannot complete its life cycle (vegetative or reproductive) without the particular element, 2) its action must be specific and cannot be replaced by another, and 3) its effect on the plant must be direct. Sixteen elements have been found essential for woody plant growth - arbitrarily divided into macronutrients and micronutrients on the basis of their normal concentrations in plant tissue.

Fowells notes that macronutrients are usually required in amounts of the order of 1 part per million or more while micronutrients are adequate when present at rates of much less than 1 part per million or as low as a few parts per billion. The macronutrients are nitrogen (N), phosphorus, (P), potassium (K), magnesium (Mg), sulfur (S), and calcium (Ca). The micronutrients are iron (Fe), manganese (Mn), copper (Cu), boron (B), zinc (Zn), molybdenum (Mo), and chlorine (Cl).

These elements are derived and absorbed from the soil. The elements carbon (C), hydrogen (H) and oxygen (O), which comprise about 90% of plant dry matter, are also needed by plants; however, carbon is obtained from carbon dioxide (CO_2) in the air, and oxygen and hydrogen from water. Soils in the Eastern U.S. support abundant growth of many forest tree species but these soils can lack P, K, Ca, Mg, S and various micronutrients.

Excessively alkaline urban soils in this region (above 7.5pH) can immobilize Fe, Cu and Zn. Potassium, S, Ca and Mo may reach toxic levels in urban soils due to high pH. Nitrogen and Mn can also become deficient in highly alkaline soils of 9.0.

Macronutrients

Nitrogen (N)

Nitrogen is the most commonly deficient nutrient in soils, particularly urban soils. Nitrogen deficiencies are also often reported in coniferous forests of cold climates under conditions that favor accumulation of thick acid humus. Incipient N deficiencies are also found in many sandy soils of warmer climates including the flatwoods and sand hills of the U.S. coastal plain and the Douglas fir region of the Pacific Northwest.

Lack of optimum nitrogen supply is probably both a cause and effect of poor tree vigor, resulting in thin crowns, yellowish-green foliage, lack of chlorophyll and progressive twig and branch die-back. Adequate nitrogen is essential for the production of amino acids, proteins, and growth hormones; it promotes vigorous growth and delays maturity. N is also an integral part of the chlorophyll molecule.

Unlike other essential nutrients nitrogen is not a product of the weathering of rock parent material. Organic matter decomposition is the prime source of nitrogen for trees. Ovington and Madgwick studied a 33 year old stand of Scotch pine in a natural forest ecosystem and found the distribution of elements between the living trees and the rest of the organic material on the site to be roughly as follows: One-half of the Ca, Mg, K and P was in the living trees, while only about one-sixth of the N was in them.

This led them to conclude that the natural N equilibrium on the site favors its accumulation in the forest humus and organic matter rather than in the trees. Soils low in organic matter, such as in many urban soils, often support trees suffering from nitrogen deficiency. Although nitrogen composes 78% by volume of the air and is one of the most abundant of the essential nutrient elements, it is largely unavailable to plants. Atmospheric nitrogen that does get incorporated into the soil occurs from the fixation of N_2 to NH_3 (anhydrous ammonia) by microorganisms.

Rainfall, because of electrical activity in the atmosphere, provides a small quantity of NH_3, as well. The NH_3 is released to the soil as organic matter after the microorganisms complete their life cycle and begin to break down Microbes further break down these complex organic forms of N into inorganic forms (ammonia and ammonium) in a process called mineralization. Mineralization must occur before plants can utilize N. N becomes available to higher plants only as the C:N ratio approaches 10:1.

Through another microbial process, nitrification, ammonium (NH_4^+) is transformed into nitrite - a transitional compound present in trace amounts - and finally into nitrate (NO_3^-); the amount of nitrate finally produced depends on the relative amounts of decomposable organic matter present. Nitrification is most efficient in well-aerated soils and at a pH of about 7.5 to 9.0 - conditions that are rare in most Eastern U.S. soils. At low pH (5.0) the rate of nitrate production from ammonium or from organic matter is slow.

Assimilation of N is more complicated than other essential elements because it is assimilated both as the NH_4^+cation and as the NO_3^- anion, and because interactions of N with other nutrients is common. Potassium increases NO_3^- uptake, while P and Cl decrease uptake of NO_3^- and enhance uptake of NH_4^+. Both NH_4^+ and NO_3^- can be removed from the soil solution by soii organisms and converted into organic N through the process of immobilization. This N is then temporarily lost to the plant until made available once again through mineralization.

Some N is volatilized into the atmosphere in the form of ammonia (NH_3) gas. This is particularly a problem in coarse-textured soils with high pH. Although plants respond more slowly to NH_4^+ than to NO_3^- there is no evidence to show that there is any difference in the eventual use to which plants put them. Kramer and Boyer state that N usually is taken up by trees as NO_3^-which is then reduced and incorporated into amino acids.

The plant will probably take up in greatest quantity whichever form predominates in the soil, although the relative amounts absorbed may be modified by the age and kind of plant, soil pH, and other environmental factors. In highly leached and acid forest soils, ammonium can be the predominant N form as a result of low soil nitrifying capacity. N becomes available to trees through: mineralization of organic matter, addition of fertilisers and fixation of N in the air by microorganisms.

N in the soil becomes unavailable to trees through absorption by grasses, weeds (plants growing where they are unwanted) and other organisms during decomposition of organic matter, volatilization into the atmosphere, and denitrification by soil organisms - a problem in waterlogged soils low in oxygen. Finally, nitrates are lost from the root zone by leaching. Climate plays a dominant role in determining the N status of soils. Within areas of uniform moisture conditions and comparable vegetation, the average N and organic matter contents of the soil decrease exponentially as the annual temperature rises.

Phosphorus (P)

Soil phosphorous can be divided into two primary classes, organics and inorganics. Organic P occurs in the form of phospholipids, nucleic acids and inositol phosphates. Most inorganic phosphorus is derived primarily from the calcium phosphates and iron and aluminum phosphates in soils, and it is believed to be absorbed by plants mostly as the primary orthophosphate ion. Phosphorus is tightly held in soils, even those that are nearly 100% sand, and its availability to plants is low. P is commonly deficient in agricultural soils but it is seldom deficient in soils in which trees and large shrubs grow.

The availability of insoluble soil P, like N, is primarily dependent on mineralization (microbial activity) in the rhizosphere. Availability of P is influenced by soil acidity, as well, as it tends to form insoluble precipitation products with Fe, Al and Mn in very acid soils. In neutral to alkaline urban soils where there is microbial activity P would tend not to be deficient. Huber reports that mycorrhizae of woody plants appear as important to P nutrition as symbiotic N fixation is for N. The surface horizons of some coastal acid sands and organic soils of the southeast U.S. are particularly low in P because of their weak capacity for P retention.

They contain very low concentrations of Fe, Al and Mn; therefore most of the P in the surface layers has been leached to lower horizons. Phosphorus provides the energy for several chemical reactions within the plant in the form of high-energy organic complexes: ATP (Adenosine Triphosphatc and ADP (Adenosine Diphosphate). Phosphorus deficiency, even though mild, will reduce this energy transfer system and slow growth functions of the plant.

Potassium (K)

Bould and Hewitt report that potassium occurs as primary and weathered minerals, and in non-exchangeable, exchangeable and water-soluble forms. The most important K-containing minerals in soil are orthoclase, microline feldspar, muscovite and the clay mineral, illite. For plant nutrition, the exchangeable and water-soluble forms are the most readily available, the nonexchangeable acting as a reserve. Most soils contain enough potassium for woody plants and trees but potassium can be deficient in soils that are acid, low in organic matter, or sandy. K is very mobile in the plant tissue, and unlike most other essential elements, it does not become a structural component of the plant.

As a regulator of enzyme activity, potassium is involved in essentially all cellular functions, including photosynthesis, phosphorylation, protein synthesis, water maintenance, reduction of nitrates, and reproduction. A balanced level of K induces thicker cell walls, accumulation of amino acids (arginine), and production of new tissues. The element tends to be concentrated in the actively growing portions of trees such as buds, current year's foliage and growing root tissues, while the proportion of K is relatively low in older, mature tissues. K plays an important role in frost hardening of trees, involving sugar-starch conversion at the end of the growing season. Its level in plants depends upon the availability of Mg and Ca and a deficiency of K impairs the utilization of P.

Calcium (Ca)

Calcium is rarely deficient since some calcium source, generally calcium carbonate or dolomite, is widely used to adjust the pH of acidic soils. In addition, most soils contain sufficient calcium for plant growth even when soil pH is relatively low. Calcium exists in soils mostly in inorganic forms, and from 50 to 1000 ppm or more may be held in an exchangeable form in the surface soil. Soils developed in regions of relatively low rainfall generally contain larger supplies of calcium than soils in humid regions.

In urban systems, calcium is also released from the degradation of sidewalks and streets and when lime is used for road and sidewalk salt in winter. Calcium has critical roles in cell division, cell development, cell wall formation, and carbohydrate movement. It complements the functions of K in maintaining cell organization, hydration and permeability. In these capacities it is involved in mitosis, enzyme activation and regulation, and membrane function.

Sulfur (S)

Sulfur occurs in rocks, especially basic igneous rocks as sulfides, e.g., the mineral pyrite, which in turn oxidize to sulfates under aerobic conditions. S is available to trees as SO_4 ions via the roots, and as SO_2 via the leaves. Although sulfur is used in approximately the same amounts as phosphorus, it is much more readily available in the soil. Sulfur-oxidizing bacteria can convert free sulfur and sulfur in organic compounds to sulfates and sulfuric acid.

Sulfur can be readily absorbed as sulfate by plants or leached from the soil in the absence of plants. Sulfur deficiency is very rare in industrial countries; rainfall, irrigation water, decomposing organic matter, the burning of fossil fuels, and fertilisers provide enough sulfur for normal plant growth in most soils. S is an essential component of three amino acids necessary for synthesis of proteins, and of two plant hormones; it is also incorporated into enzymes and vitamins.

Magnesium (Mg)

Magnesium is the only mineral constituent of chlorophyll and is also associated with rapid growth, carbohydrate metabolism and oxidative phosphorylation in young plant cells. Mg tends to be deficient and readily leached in sandy, acid soils. P and Mg are often deficient in the same soils and they can be antagonistic. Mg occurs in several minerals, e.g., micas, hornblende, dolomite, serpentine and montmorillonite.

Micronutrients

Most soils contain sufficient amounts of micronutrients to promote tree growth. Deficiencies do develop, however, and are found most often in soils outside a pH range of 6.0 - 7.0 or in sandy, well drained soils where heavy rainfall encourages leaching. Iron and manganese are the micronutrients most frequently found deficient even though these nutrients are present in the soil in concentrations adequate to support growth. The situation is usually corrected by adjusting the pH, which removes the elements from fixed, insoluble compounds and renders them available for root uptake.

Iron (Fe)

Harris states that iron deficiency is the most common micronutrient deficiency. Fe deficiency is common on many species grown in areas of low rainfall and in alkaline soils (pH above 7.0) where high levels of calcium tie up the iron in insoluble forms - its deficiency is sometimes called lime-induced chlorosis. In many studies, chlorosis of pin oak, white oak and red maple have been reduced or eliminated by surface and subsoil application, as well as, trunk injection of sulfur.

It is not uncommon to find Fe deficiencies in many urban soils. High levels of P, either alone or in conjunction with calcium, may also tie up iron in insoluble complexes. The ferrous forms of iron are the most available for plant nutrition. Iron is essential for chlorophyll and the reactions of photosynthesis and plays a role in the synthesis of proteins and the function of certain enzymes.

Manganese (Mn)

Manganese reacts similarly to iron in many respects - it is essential in the synthesis of chlorophyll and is similarly affected by calcium and phosphorus. Manganese deficiency is more likely to occur in poorly drained soils that are high in organic matter. The temperature of soils affects its solubility. As the temperature of soils decreases, so does manganese solubility.

It could potentially be deficient in alkaline urban soils; the solubility of manganese decreases as soil alkalinity increases, and it is not readily available to plants above pH 6.5. Manganese is a constituent of only one known plant component, but it activates various enzymes involved in nitrate reduction, carbohydrate metabolism, and respiration. At high concentrations, Mn competes with Fe for absorption and translocation.

Copper (Cu)

Field experiments have shown that copper sulfide acts as a source of copper for plant growth - copper sulfide probably originated from the most important copper compound in primary rocks, chalcopyrite. There is some evidence that atmospheric sources may provide significant amounts of Cu, as well. Copper deficiency is fairly widespread and is not likely to occur on soils that are sandy, organic, alkaline or calcareous.

Copper deficiency can be aggravated by alkaline irrigation water and by nitrogen or phosphorus accumulation - conditions that generally would only be encountered in the arid or semi-arid regions of the western United States. Foliar and surface application of copper fertilisers can successfully eliminate deficiencies. Copper is toxic to cambium and sapwood and trunk injections are, therefore, not recommended. Copper is a component of several enzymes and is involved in protein and carbohydrate synthesis, and N fixation.

Boron (B)

There is a relatively narrow range of concentration between deficiency and phytotoxicity of B. According to Whitcomb boron deficiency in soils is not as common as boron toxicity. It is required by plants in very small amounts and functions in translocation, cellular differentiation and development, carbohydrate metabolism, and the uptake or translocation of Ca. In the United States, areas of known boron deficiency are located in the eastern third of the country and portions of the Pacific states. In one California county, boron-deficient soils are within 6 miles of soils with excess boron.

Molybdenum (Mo)

Molybdenum is also required in very small quantities in plants, but nonetheless is very essential for the transformation of nitrate into amino acids. In contrast to most of the other micronutrients, Mo is less available at low pH; Mo deficiency commonly occurs in soils that severely lack phosphorus and sulfur.

Chlorine (Cl)

Chlorine is the only essential element for which a deficiency has not been observed under field conditions. Excess chlorine is much more of a concern, particularly in irrigated arid regions and near seacoasts. Trees adjacent to roadways treated with salt during the winter can experience serious injury, as well, thus the symptomology of cloride injured trees needs to be kept in mind in northern and eastern U.S. urban areas where calcium chloride is used for winter de-icing on sidewalks and streets.

Zinc (Zn)

Zinc deficiency is common among cultivated trees and large Zinc compounds decreases in solubility as pH increases and, like many other nutrients, is more likely to be unavailable in the soil than low in total quantity. Rarely would Zn deficiencies be encountered in urban areas. The primary physiological role for Zn is its interrelationship with auxin. Addition of Zn to deficient plants greatly stimulates auxin synthesis - thereby making it essential for cell elongation and growth.

NUTRIENT DEFICIENCY AND NUTRIENT REQUIREMENTS

A fertiliser program should be used in urban systems when there is a documented nutrient deficiency and the nutrient requirements of the tree cannot be supplied in adequate amounts by the soil or when there is occasion to believe that our objectives would be better served by the application of fertiliser. It should take into consideration environmental factors, which can make added nutrients unavailable such as the chemistry and physical characteristics of the soil and amount and intensity of precipitation.

WATER MANAGEMENT

Urban trees probably suffer more from water-related problems than from any other cause and urban foresters would be well advised to use irrigation first in attempting to correct a tree's health problem. It is often difficult, however, to distinguish between the effects of nutrient deficiency or drought. Nutrient deficiencies limit root growth and make plants more susceptible to drought; similarly, moisture deficits will restrict nutrient uptake in soil. Many factors exacerbate the water problem in the urban environment. Water is added to the soil primarily through rainfall, but much of this essential water is lost either as runoff over impervious paved surfaces, through drainage beyond the reach of roots or as evaporation from the soil surface.

We have already established that the many reflective and absorptive surfaces in the urban environment are responsible for an increase in daytime and nighttime temperatures, for drying the air and creating a "heat island" which increases the tree's water needs. If water is scarce, tree growth will be slow. Under extreme conditions, the tree may die. Even more importantly measurable growth increases were evident at the end of the season. Along with irrigation, restricting unnecessary evapotranspiration losses from the soil using weed control and mulch techniques can increase the supply of water in the soil.

Following weed control, response to added fertiliser may be small. Not only are the weeds no longer competing for soil reserves of water and nutrients, but there is a flush of released nutrients from the killed weed biomass, either or both of which may make additional fertiliser unnecessary on reasonable soils. On industrial or derelict urban soils there may be a demonstrable need for the addition of nutrients. If this is the case, recommendations for fertilisation should specify:

1) the appropriate timing of fertilisation,
2) the right form and amount of fertiliser, and
3) the appropriate method or methods of application.

Timing of Nutrient Application

To make the best use of fertilisers an arborist must know when trees benefit most from their application. The timing of nutrient application is determined and affected by many factors including: climate, rainfall, soil type, and plant species and their associated physiological and developmental states. Most of the current literature recommends application of nutrients in either the spring or fall. Leaching could be a potential hazard with mid-fall applications of fertilisers in colder northern or northeast U.S. soils where heavy precipitation coincides with slow or ceased root activity. In southern or west coast climates where the temperatures are mild the year round, application should be made before the growing season starts. As long as soil temperatures are above 40-45 F, roots can absorb nutrients. An added advantage is that in spring and mid-fall, soil moisture conditions also favor plant nutrient uptake.

Methods for Nutrient Application

The primary methods of applying nutrients that are commercially practiced today include: surface broadcasting, deep-root feeding (also called punch bar system, auger holes or soil drill holes), liquid soil injection, foliar spraying and injection or placement of nutrients in holes (implants) in tree trunks. Other methods used, more appropriate for orchards and plantations, include soil incorporation by rototilling or discing and addition to irrigation water. Each application method serves a specific role depending on the site and plant condition. The appropriate method will depend on the types of fertiliser used, the specific objective of fertilising, costs, the equipment available, other plants in the vicinity, nature and slope of the soil surface, and, in certain cases the species to be treated.

Surface application

The broadcast application of fertiliser to the soil surface was one of the first methods employed and is the easiest and least expensive method of applying N, most micronutrients, and, where roots are shallow. It requires the least amount of time and does not require sophisticated equipment. The fertiliser is broadcast over the soil surface using a spreader that is calibrated to apply the desired amount of N per 1000 sq. ft. Suggested rates of application for homeowners range from 1 to 3 lbs. of actual N/1000 sq. ft. It is generally recommended that an area that extends from the trunk to the dripline and a third of that distance beyond the dripline be covered. Use of surface applications are most appropriate in trees and shrubs in open beds, and where water penetration is not limited by compaction or excessive slopes. On compacted soils, and, when possible he recommends raking the area or using other scarifying methods to loosen the soil.

There are some objections to the surface method of applying nutrients. On lawns, some professionals believe that turfgrass will significantly compete for available N and turf in fertilised areas will be much greener than in surrounding areas. Turf competition can influence nitrogen availability and tree growth. When turf limits only the nutrient availability - without affecting water availability - applied nitrogen will increase tree growth. However, excess nitrogen applied to trees can slow growth and create stress. Although relatively little is known about the nutritional needs of urban trees, their requirements are generally less than those of turfgrass species. Thus, fertiliser regimes designed for turf may impact tree health.

Burning of grass or injury to roots from excessive surface fertiliser application is, also, often cited as a disadvantage of this method of application. Fertiliser burn results when excessive soluble fertiliser in the root zone increases the salt concentration, and hence, the osmotic pressure of the soil solution. This can prevent water from entering the root cell and may even cause water to move out of the cell. When this occurs the root cell dies and may trigger the death of the entire plant. Fertiliser burn is most problematic when soil moisture is limited; often at the driest part of the summer months.

Adding excessive amounts of fertiliser acidifies soils and consequently could cause immobilization or toxicity of other essential nutrients as well as inhibit soil microbial activity - not to mention the increased potential for excessive pollution from leaching and runoff. Some researchers and others recommend generous watering to dissolve the soluble salts, dilute them, and wash them down into the soil to avoid fertiliser burning of grass and injury to tree roots. Applying the fertiliser in smaller quantities, dividing it into two or more applications at different times, and spreading the fertiliser uniformly may also reduce the hazard of injury from salt concentrations but is much more costly. There is evidence that no significant benefits are derived from nutrient application techniques other than simple broadcast surface techniques.

Various fertiliser formulations and four methods of application were used: surface broadcasting, placement of dry fertilisers in holes, injection of liquid fertilisers into the soil, and foliar spraying. Statistically the results from foliar treatments were not significantly better than the results from no treatment.

All three methods of applying fertiliser to the soil stimulated tree growth. The three methods of soil application appeared to be equally effective, with minor variations among the tree species.

Drill hole or Auger method

Placement of fertilisers in auger or drill holes is a common practice by commercial arborists. It is more costly than surface broadcasting but may be the preferred application method under certain conditions. One of the main benefits is encouragement of deeper rooting by improving subsoil fertility. Where lateralrooting space is limited, or soils cause superficial rooting, these procedures may be therapeutic. These methods are useful only in soil where tree roots are not near the surface, such as in bare or cultivated soil. It is under these circumstances that P, K and other nutrients of low solubility usually need to be incorporated into the soil, closer to the root zone where they will be more readily available. The drill hole method of fertilisation also places fertiliser below turfgrass roots, avoiding fertiliser burn and providing less competition with turf roots for nutrients. The soil is aerated and water penetration increased by this method, as well. It is especially beneficial for opening up heavily compacted soils.

Liquid injections

Liquid formulations (fertilisers suspended in water) or any soluble dry fertiliser may be injected into the soil with water using a lance and hydraulic sprayer. The advantages to this method are the better distribution and rapid uptake of fertiliser and when available water is the factor limiting fertiliser absorption by roots liquid injection reduces this problem. Smith recommends this method when deficiencies are readily apparent provided that the soil physical characteristics allow for adequate nutrient movement in solution.

Foliar nutrient levels were lower than optimum and the trees were considered to be macronutrient deficient. Two application methods of complete fertilisers were compared: trunk injection and liquid injection below the soil surface. In their study, soil injection of the slowrelease fertiliser increased the foliar nitrogen level and improved color more than two applications of trunk injected macronutrients. There are several disadvantages to the soil injection method that require consideration. Fertilisers must be soluble to be used in this method. Water-soluble fertilisers containing P, K and other nutrients are usually more expensive than the less soluble compounds making this method rather costly. The procedure is also time consuming (therefore, more costly) and fertiliser solutions are notoriously corrosive to equipment. The effects of liquid injection may be less persistent than that of dry fertiliser and if the soil will not absorb the large amounts of water the possibility of runoff and leaching exist.

Foliar sprays

Although foliar sprays can be used for macronutrients (N, P, K, Mg, Ca), which are subject to immobilization and leaching, they are best used as a rapid way to overcome deficiency symptoms involving micronutrients like iron, zinc, manganese, copper and boron. Micronutrients are required in extremely small amounts and can usually be absorbed quickly through the leaves, so deficiencies can usually be quickly and safely corrected by foliar sprays. Foliar application is particularly effective in treating iron chlorosis. Small-scale spray applications of a single micronutrient may quickly determine a plant's response, which can be a very useful diagnostic tool to plan a longer corrective treatment program.

In some commercial practices foliar spray material is also used to help get young and recently transplanted trees established in the landscape. Micronutrient spray applications are most effective when made just before a period of active growth; but that not all plants respond to foliar treatments. Sprayed materials may penetrate young leaves more readily than older leaves and the chelated forms of micronutrients are generally absorbed well. Temperature and spray concentration can also affect the tree's ability to take in the sprayed materials; high temperatures and low humidity reduce absorption. A way to allay the expense of foliar sprays is to include pesticides in the formulation. Although this is being incorporated in many horticultural crop spray programs, it is not being done routinely on landscape plants. Large trees and shrubs are often difficult to spray and the danger of involving non-targeted plants, people and property exists.

Implants and injections

Implants (such as capsules) and injections (micro- or macro-injections) are two techniques, which introduce chemicals directly into the xylem - relying on the transpirational stream to move materials systemically through the tree. Like foliar sprays, these methods are most often used for the correction of micronutrient deficiencies. They have also been used to treat certain pest problems. Some authorities are hesitant to recommend this method because it has the potential to provide easy access for fungi diseases and rot, and these fertilisation methods should only be used when other methods are too difficult or have failed.

Numerous holes must be routinely drilled (every 3 to 4 years) into the bark which can lead to: excessive sap leakage, toxicity to the cambium and xylem where salts are inserted and toxicity to foliage when too much nutrient is applied at the incorrect time to sensitive plants. In all the control wounds they found negligible amounts of discolored wood and cambial dieback indicating that the wound made by the injection tube is not serious. Similar holes that were injected with nutrients were more seriously affected.

Fertiliser Application Rates

Discussion of fertiliser application rates and later discussion of types of fertilisers will focus on the fertiliser element nitrogen. Of all the nutrients N has the greatest effect on plant growth, is almost always deficient, is required in the largest quantity by the plant, and generally is the most difficult nutrient element to manage. In fact, it is often the only nutritional element that accelerates growth in young shade trees. Nitrogen deficiency and management are major issues in disturbed urban soils under stressful urban micro-climatic conditions. Recommendations and rates of application for all nutrients except N must be based on a soil test or analysis of the foliage. The application rate of N depends on several factors including species, stage of development of the plant, location and formulation.

Environmental factors which make N and any other added nutrient, for that matter, unavailable or encourage volatilization and leaching losses should also be considered. Precipitation amounts and distribution, natural fertility, chemistry and physical structure of the soil should all be given careful attention, as well. Whitcomb states that rate of N application is more influenced by the porosity of the soil and leaching of the N below the root zone than by any other factors. Heavy clay or silt soils may require limited N, whereas sandy soils require more frequent application and, in extremely sandy soils slow-release N sources are most beneficial. Rates of actual N are usually expressed in lbs. per acre or in lbs. per 1000 sq. ft.

Fertiliser Dosages

Some authorities and arborists relate the amount of fertiliser to be applied to trunk diameter - usually 2 lbs. of complete fertiliser for every inch of trunk diameter. However, fertiliser rates based on root zone are considered a better practice than rates based on trunk diameter and application rates based on the latter practice are becoming obsolete. There is no functional relationship between root spread and trunk diameter and points out that by using the trunk diameter method application rates actually drop as larger trees are fertilised. For example, a tree with a 2-inch trunk would receive 4lbs. of fertiliser; a six inch tree would get 12lbs. But even if the size of the root system was related to the trunk diameter, this method would reduce the rate of fertilisation. While the diameter of the root system would be tripled the area of the root zone would increase 9 times; so in fact, the area would need 9 times more fertiliser, or 36 lbs. instead of 12.

Application rates of fertiliser based on area of root zone appear most appropriate in light of research done on fibrous root networks of shade trees showing that roots extend far beyond the tree's crown or dripline. Perry citing work by Stout and Lyford and Wilson reports that it is not uncommon to find trees with roots systems having an area with a diameter one, two or more times the height of the tree. As stated earlier, the area to be fertilised is governed by the location of the absorptive roots which for most trees lie mainly in the outer two-thirds of the circular area around the tree. Trees growing in sandy well-drained soils have more extensive roots systems than those in finer texture soils and root extension also varies greatly among species. Roots of narrow or columnar trees extend well beyond twice the crown radius. The root area of sugar maples can be 1.75 times larger than the crown area and root area of tulip poplar 2.5 times larger, whereas for pin oak area is about the same as crown area.

There are other similar methods to calculate application rates of complete fertilisers that use the square footage of the circular or rectangular area under the tree's canopy. Fertiliser rates and dosages, for the most part, appear to be based less on empirical evidence and experimentation than on anecdotal impression and general "common sense". None of these calculation methods addresses itself to the urban environment where the roots of the tree are frequently overlooked and disrupted. The small volumes of compacted soil that the roots have access to are either poorly drained, or more likely, cannot hold enough water to meet demand, and the trees experience periodic to prolonged drought. In other words, lack of soil volume accounts for most urban tree survival problems and fertilisation could be more injurious than beneficial under these conditions.

Recommended Amounts of N Fertiliser

Research at the University of Tennessee, Knoxville, has shown 150 lbs. actual N per acre (3.4 lbs. N/1000 sq. ft) greatly accelerates the growth of established young shade trees. Wright and Hale experimenting on the influence of N rates on growth of red maple, pin oak and flowering dogwood at Virginia Polytechnic Institute and State University, demonstrated that a rate of 300 lbs. N per acre (6.8 lbs. N/1000 sq. ft) was wasteful after the third year because it did not significantly increase growth beyond that obtained with 150 lbs. per acre (3.4 lbs. N/ 1000 sq. ft) In general, for most soils, recommendations range from 2 - 6 lbs. of N per 1000 sq. ft (about 90 to 260 lbs. per acre) per year. Pirone et al. advocate applying rates of 2 - 4 lbs. of N/1000 sq. ft. Hamilton et al concur and recommend these yearly amounts be divided into two or more applications during the growing season. 3 - 6 lbs. of N/1000 sq. ft ; 3 to 4 lbs. of P and 6 lbs. of K per 1000 sq. ft are recommended by Williams for woody plant growth. For established landscape trees, Doughty considers 1 - 2 lbs. of actual N/1000 sq. ft adequate.

For infertile soils or subsoil materials exposed at the surface Mader and Cook advise incorporation of a multi-nutrient (N-P-K) fertiliser such as 10-6-4 at a rate of 25 - 50 lbs/1000 sq. ft (2.5 - 5 lbs. N/1000 sq. ft). Neely and Himelick state that the need and frequency of application of P and K should be determined by chemical tests. If needed, these nutrients may be added at intervals of 3 - 5 years at the following rates: P at 2.5 - 3.6 lbs/1000 sq. ft and K at 5 - 6 lbs/1000 sq. ft. K remains readily available and cycles from tree to soil for many years, so unless leached away or removed in leaves one application will last for several years.

When turf is present, Hamilton et al. and Doughty advise applying a concentration of no more than 2 lbs. total N per 1000 sq. ft. The total yearly amount of fertiliser should be calculated to include the amount that is applied for the turf grass. If the lawn is already being fertilised, it is unlikely that the tree will need much more fertiliser or any at all. Koelling and Kielbaso recommend fertilising trees surrounded by turf using the drill hole method to avoid harming the grass. They recommend 6 lbs. N/1000 ft applied in 3 applications at 2 week intervals for trees (with grass growing beneath) less than 25 ft high; and 6lbs N/1000 sq. ft in one application for trees greater than 25 feet high. They discourage fertilisation of ornamental flowering trees and other small fruit trees unless a definite needs exists. They maintain that heavy applications of N may tend to reduce flowering. Again, further research needs to be undertaken specifically on urban trees which often grow in limited spaces under unique and adverse conditions.

Fertilising at Planting and Transplanting

Sodium nitrate, which is almost never used today, has one of the highest salt indexes of any fertiliser material. The higher the salt index, the greater the likelihood that the plant will experience water stress or fertiliser burn. Despite other sources of N, placement of fertilisers in planting holes is still not recommended by some professionals. Fertilisation of transplanted trees is also disputed. Much of the controversy focuses on the concept of top (or shoot): root ratio. The shoot:root system has a much larger top to support in terms of water and nutrients (which may increase stress on the tree during periods of low precipitation) and a root system that may be disproportionately small to provide adequate anchorage for tree.

Only a small percentage of the original root system is moved with a transplanted tree. Rapid root regeneration and adequate soil moisture are the most important factors for the successful establishment of transplanted trees.

Benefits of Fertilisation

Depending on the objectives, fertilisation may benefit established trees in a variety of ways. Proebsting fertilised plots of almond trees with ammonium sulfate and increased almond yields 175 percent. Fertilising stimulated diameter and height growth of stands of eastern cottonwood, sweet gum (Liquidambar styraciflua L), water and willow oak, tulip poplar and sycamore in studies conducted on seedlings and 20 year old trees by Broadfoot and Broadfoot and Ike.

Doubling the rate of application did not double the increase in growth and was slightly injurious to the tree. At even higher rates (10 to 20 lbs. N per tree) the fertiliser was so injurious that earlier effects of increased growth and vigor were entirely obscured. General appearance of treated trees was superior to that of untreated trees and analyses of leaves from a number of plane trees and pin oaks showed more of both N and P. Not all trees under all soil conditions respond to fertilisation, however.

One application, each, of three levels of ammonium nitrate was added and data were collected for the next 28 months. N fertiliser did not affect either foliar nitrogen concentrations or tree diameter growth. From these data, it appears that the native fertility of the soil provided adequate nutrition. Age of the tree appears to be a factor in its response to fertilisation. Smith and Treaster studied the effects of fertilisation on linden, apple trees and sugar maples and found that 15 years after planting there were no differences in growth increases (trunk caliper, height or branch diameter) between the control and treated trees.

Types of Fertilisers

There are over 2000 grades of fertilisers which contain N in a variety of formulations - these include liquid, granular, sulfur coated, and encapsulated. Nitrogen fertilisers come in different chemical forms but are absorbed by plant roots almost entirely as ammonium and/or nitrate; the most common of the inorganic forms. As mentioned earlier, NH_4^+ and NO_3^- have important chemical differences that affect their availability. Generally, the nitrate anions are mobile in the soil, while the ammonium cations initially react with clay and other colloidal surfaces, remaining relatively immobile until nitrified by soil organisms.

The long-term effect of ammonium fertilisers is to reduce soil pH. In the conversion of NH_4^+ to NO_3^-, an acid residue forms, lowering soil pH. This may lead to formation of an "acid roof" where the pH in the upper inch or two of soil may be as much as an entire pH unit less than the soil beneath. Applying limestone may counter balance this effect and decrease chances of root and turf injury.

Organic or Inorganic Fertilisers

Fertilisers are available in both an organic or inorganic form. Some of the soluble inorganic nitrogen fertilisers include: ammonium nitrate (33% N), calcium nitrate (15% N), potassium nitrate (14% N, 44% potassium), monoammonium phosphate (12% nitrogen, 48% phosphorus), sodium nitrate (16% N), and ammonium sulfate (20% nitrogen). Most liquid nitrogen fertilisation is done with ammonium nitrate and potassium nitrate. Calcium nitrate is sometimes used as a foliar spray but seldom for soil applications, due to the much higher cost. Organic fertilisers are carbon-based and can be natural (e.g. manures, sewage sludge, bone meal) or synthetic (e.g. isobutylidene diurea (IBDU) or ureaformaldehyde).

Organic fertilisers also release inorganic ions, but more slowly as the molecules are hydrolyzed or decomposed in the soil, thus reducing losses through leaching. The solubility of inorganic fertilisers, however, is less affected by temperature so that the rate of availability is more uniform. When nutrients are the primary interest, inorganic forms are usually favored. The principal advantage of natural organic fertilisers is that they improve soil tilth (structure) if incorporated into surface soils. Urea is technically an organic, but it is not recognized as such in the industry because of how rapidly it solubilizes and releases its nutrient ions in water.

Urea has the highest nitrogen concentration (46%) and is generally the least expensive of the granular fertilisers. However, urea provides only ammonium nitrogen initially and under cool conditions, the conversion to nitrate is very slow. This may be an advantage or disadvantage depending on the immediate plant needs and the species involved

Slow-release Fertilisers

A justifiable belief is that control of the rate of solution, ammoniafication and nitrification of a nitrogen source will result in increased efficiency of nitrogen use. Most of the factors contributing to loss of efficiency are directly related to rapid dissolution and hydrolysis of the applied compound, i.e., N loss through leaching, denitrification of nitrate, and ammonia volatilization. One approach to increase efficient use of nitrogen has been to develop fertilisers that release N slowly into the soil solution, over an extended period of time. These compounds are referred to as controlled release, slowly soluble or slow-release fertilisers.

The production of slow-release N fertilisers involves either compounds that have reduced dissolution rates or compounds that are encapsulated with some material which delays or slows the rate of release. Examples of the first type are the metal ammonium phosphates, oximide, and the ureaforms, such as isobutylidene diurea and ureaformaldehyde. The second type of slow-release N fertiliser is attained by such coatings as waxes, parafins, acrylic resins, latex, gums, oils and perforated plastic. Hauck reports that an effective coating has been developed by TVA (Tennessee Valley Authority) that uses a low-cost material, sulfur, sealed with a thin film of petroleum wax and a microbicide. Applying fertilisers in the slow-release form has several advantages in urban environments.

Fertiliser does not need to be applied as frequently if applied in a slow-release form and higher amounts of fertiliser can be used without raising salt levels enough to injure plant roots. In urban soils where water deficits are the norm, applying slow-release fertilisers may reduce the risk of salt injury to roots. The primary disadvantage of slow-release fertilisers is the higher cost. They range from two to six times higher per unit of N than ammonia N before application; therefore their use must be considered on the basis of demonstrated increased efficiency.

It is also important to emphasize that slow-release sources of all elements should be evaluated carefully to determine not only the rate to be used but how much of the nutrient is released over what period of time. Slow-release fertilisers are especially useful on salt sensitive plants and Whitcomb maintains that, given all other factors are equal, a slow-release source of nutrients will provide for plants more efficiently than dry chemical fertilisers. May and Posey reported equivalent growth of pine seedlings either from either a single application of 248 lbs. of ureaform N/acre or 376 lbs. of N/acre from ammonium nitrate distributed over 8 applications.

Nitrification inhibitors

Another strategy to reduce the loss of N from applied fertiliser is to slow the conversion of NH_4^+ to NO_3^- (nitrification) by inhibiting the activity of the bacteria responsible for the process. This concept is presently being used in grain crop production and on citrus trees. Serna experimenting with citrus trees (Citrus spp.) found that nitrification inhibitors such as dicyandiamide (DCD) helped to reduce leaching losses by retaining applied N in the ammonium form. Adding DCD to trees receiving an ammonium sulfate-nitrate fertiliser (ASN) resulted in higher N concentrations in the spring flush leaves, higher number of fruits per tree, and better fruit color index than trees treated with ASN alone. These results suggest that the use of a nitrification inhibitor permitted a more efficient utilization of fertiliser N by citrus trees. Nitrification inhibitors could be of possible value in sandy urban soils and urban soils where leaching is a problem.

Phosphorus and Potassium Fertilisers

Phosphorus and potassium are rarely in short supply for trees, are deficient in fairly specific soils and overuse of fertilisers containing these elements may lead to toxicity symptoms on plants and to water pollution. High concentrations of phosphorus (and nitrogen) increase plant life in lakes and streams which may result in low oxygen levels that may be fatal to fish. The loss of phosphorus is almost completely due to surface soil erosion, though it can be leached from coarse-textured or sandy soils. The predominant soil type of the southeastern U.S., Ultisols, is particularly prone to erosion and fertilisation with P should be undertaken only when there is a documented need and then only with caution.

Gowans advises against using phosphate to fertilise trees on sites that are susceptible to water runoff, erosion or leaching. If P is necessary for growth or health of the tree the amount used and the method of application should be carefully controlled to minimize transport of phosphate from the site. When P is not needed application will increase soil salinity and could tie up micronutrients - especially zinc and copper. P and K are largely immobile in soil; therefore incorporating them into the soil as opposed to applying them on the soil surface is recommended. Phosphorus fertiliser materials currently available range from raw rock phosphate to higher condensed forms of polyphosphates. Rock phosphate has been used successfully with (Pinus elliottii Engelm.) slash pine and radiata pine. With slash pine it was found that rock phosphate was an effective P source on flatwood sands but not on other soils tested.

Common N and K fertilisers are all water-soluble, but P products cover the range from zero to high levels of citrate or water solubility. The most common P fertilisers used are the slow-release, water-soluble superphosphates. The water-insoluble P fertilisers include basic slag, dicalcium and tricalcium phosphate. There are also mixtures of water-soluble and water-insoluble P fertilisers. There are several forms of potassium fertiliser. Potassium chloride (KCl) is very soluble and has the highest salt level of any fertiliser material. It should not be used where salinity is a problem and in arid regions where chloride may be toxic. Potassium sulfate (K_2SO_4) is preferred for most uses. Potassium nitrate (KNO_3) is frequently used in liquid fertiliser systems in combination with ammonium nitrate but may cause N excess problems if used to supply large quantities of potassium.

Complete Fertilisers

A complete fertiliser is one that contains significant amounts of the three primary nutrients; N, P and K. A fertiliser analysis, the relative percentages of these three nutrients, is listed on the label and referred to as the N-P-K number. Before addressing the use of complete fertilisers it is important to point out, again, that the effectiveness of any fertiliser and nutrient source will be directly dependent upon the moisture conditions of the site. Lack of soil moisture not only results in moisture deficiency for trees but also interferes with nutrient absorption. Readily soluble fertiliser sources were most efficient under moderate moisture levels while under conditions of high precipitation, slowly soluble forms were more effective. Survival of longleaf pine seedlings after a drought was greatly increased by a complete fertiliser, but not by N alone.

Some researchers find that, either too much or too little N resulted in intensified drought damage and decreased post drought recovery in loblolly pine seedlings. These studies show that the value of any fertiliser is distinctly affected by the amount of precipitation and soil moisture, first and foremost. Some others believed that even though scores of field experiments have demonstrated that most soils contain sufficient amounts of P and K for trees and woody plants, complete N-P-K fertilisers are still widely endorsed for trees. These recommendations for shade tree fertilisation are derived from experience with field crops - where N, P and K are commonly deficient.

Root crops have shown increased growth with high supplements of K and grain crops often benefit from the addition of P. Primarily, one could conclude because the crop is harvested each year and the nutrient removed from the soil. This is not the case with urban trees and therefore the logic behind using a complete fertiliser in urban environments, without demonstrated need for the element, should be seriously questioned.

The indiscriminate use of complete fertilisers may lead to a nutrient imbalance that will be difficult to overcome. The effect of high K to N ratios on plant growth and evidence that there is little or no growth increase of shade trees in response to nutrient addition other than N leads to the conclusion that for increase of growth rate of shade trees, a high level of N combined with relatively low levels of P and K is most effective.

Chapter 10

Soil Testing and Fertility

Soil testing is the single most important guide to the profitable application of fertiliser and lime. Soil tests are used to monitor the impact of past fertility practices on changes in a field's nutrient status. To accomplish this, one must (1) collect samples to the proper depth; (2) collect enough samples per unit of land area; (3) collect samples from precisely the same areas of the field that were sampled in the past; and (4) collect samples at the proper time.

Depth of sampling: The proper sampling depth for pH, phosphorus, and potassium is 7 inches. For fields in which reduced-tillage systems have been used, proper sampling depth is especially important, as these systems result in less thorough mixing of lime and fertiliser than a tillage system that includes a moldboard plow. This stratification of nutrients has not adversely affected crop yield, but misleading soil test results may be obtained if samples are not taken to the proper depth.

Under reduced-tillage systems, it is important to monitor surface soil pH by collecting samples to a depth of 2 inches from at least three areas in a 40-acre field. These areas should represent the low, intermediate, and high ground of the field. If surface soil pH is too high or too low, the efficacy of some herbicides and other chemical reactions may be affected. Number of samples per unit of land area. The number of soil samples taken from a field is a compromise between what should be done (information) and what can be done (cost). Sampling at the rate of one composite from each 21/2-acre area is suggested.

Field sampling studies show large differences of soil test levels in short distances in some fields. If you can use computerized spreading techniques and suspect large variations in test values over a short distance, collecting one sample from each 1.1-acre area will provide a better representation of the actual field variability. The increased sampling intensity will increase cost of the base information but allows for more complete use of technology in mapping soil fertility patterns and thus more appropriate fertiliser application rates.

The most common mistake is taking too few samples to represent a field adequately. Taking shortcuts in sampling may produce unreliable results and lead to higher fertiliser costs, lower returns, or both. Precise sample locations. Since test results may vary markedly in short distances, it is important to collect soil samples from precisely the same points each time the field is tested. This practice reduces the variation often observed between sampling times. Sample locations may be identified using global positioning system (GPS) equipment or by accurately measuring the sample points with a device such as a measuring wheel. Once locations have been identified, collect and composite five soil core samples 1 inch in diameter to a 7-inch depth from within a 10-foot radius around each point.

How to sample: A soil tube is the best implement for taking soil samples, but an auger or a spade also can be used. Five soil cores taken with a tube will give a satisfactory composite sample of about 1 to 2 cups.

When to sample: Sampling every 4 years is strongly suggested. To improve the consistency of results, samples should be collected at the same time of year. Sampling done within a few months of lime or fertiliser treatment will be more variable than after a year.

Late summer and fall are the best seasons for collecting soil samples because potassium test results are most reliable during these times. The potassium soil test tends to be cyclic, with low test levels in late summer and early fall and high test levels in late January and early February.

Testing soil to predict the need for nitrogen fertiliser is complicated by the fact that nitrogen availability-both the release from soil organic matter and the loss by leaching and denitrification · is regulated by unpredictable climatic conditions. Under excessively wet conditions, both soil and fertiliser nitrogen may be lost by denitrification or leaching. Under dry conditions, the amount of nitrogen released from organic matter is low, but under ideal moisture conditions, it is high. Use of the organic-matter test as a nitrogen soil test, however, may be misleading and result in underfertilisation. Attempts to identify new nitrogen soil tests have been ongoing for several years.

Tests are available for most secondary nutrients and micronutrients, but interpretation of these tests is less reliable than tests for lime, phosphorus, and potassium. Complete field history and soil information are especially important in interpreting results. Even though these tests are less reliable, they may be useful in two ways:

1. Troubleshooting (diagnosing symptoms of abnormal growth). Paired samples representing areas of good and poor growth are needed for analyses.
2. "Hidden-hunger checkup" (identifying deficiencies before symptoms appear). Soil tests are of little value in indicating marginal levels of secondary nutrients and micronutrients when crop growth is apparently normal. For this purpose, plant analysis may yield more information.

Soil test ratings (given in Table 1) have been developed to put into perspective the reliability, usefulness, and cost-effectiveness of soil tests as a basis for planning a soil fertility and liming program for Illinois field crops. Additional research will undoubtedly improve some test ratings. Interpretation of soil tests and formulation of soil treatment program. Formulate a soil treatment program by preparing field soil test maps to observe areas of similar test levels that will benefit from similar treatment. Areas with differences in soil test pH of 0.2 unit, phosphorus test of 10, and potassium test of 30 are reasonable to designate for separate treatment.

When the soil test is variable. When there is large variation among tests on a field, the reason and, more important, what to do about it may not be obvious. First look at the pattern of the tests over the field. If there is a definite pattern of high tests in one part and low in another, check to see whether there is a difference in soil type. Second, try to recall whether the area was farmed as separate fields in the recent past. Third, check records for this field from previous tests or, if there are no records, try to remember whether portions were ever limed or fertilised differently during the past 5 to 10 years.

Whether or not the explanation for large differences in tests is found, split the field and apply basic treatments of lime and fertiliser according to need. If there is no consistent pattern of high and low tests, select the median test, which is the test that falls in the middle of a ranking of tests from the area from low to high. If no explanation for large differences in tests is found, consider taking a new set of samples.

Cation-exchange capacity: Chemical elements exist in solution as cations (positively charged ions) or anions (negatively charged ions). In the soil solution, the plant nutrients hydrogen (H), calcium (Ca), magnesium (Mg), potassium (K), ammonium (NH_4), iron (Fe), manganese (Mn), zinc (Zn), and copper (Cu) exist as cations. The same is true for nonplant nutrients such as sodium (Na); barium (Ba); and metals of environmental concern, including mercury (Hg), cadmium (Cd), chromium (Cr), and others. Cation-exchange capacity is a measure of the amount of attraction for the soil with these chemical elements.

Table 1: Ratings of Soil Tests

Test	Rating
Water pH	100
Salt pH	30
Buffer pH	30
Exchangeable H	10
Phosphorus	85
Potassium	60
Boron (alfalfa)	60
Boron (corn and soybeans)	10
Iron (pH > 7.5)	30
Iron (pH < 7.5)	10
Organic matter	75
Calcium	40
Magnesium	40
Cation-exchange capacity	60
Sulfur	40
Zinc	45
Manganese (pH > 7.5)	40
Manganese (pH < 7.5)	10
Copper (organic soils)	20
Copper (mineral soils)	5

In soil, a high cation-exchange capacity is desirable, but not necessary, for high crop yields, as it is not a direct determining factor for yield. Cation-exchange capacity in soil arises from negatively charged electrostatic charges in minerals and organic matter. Depending on the amount of clay and humus, soil types have a characteristic amount of cation exchange. Sandy soils have up to 4 milliequivalent (meq) per 100 grams of soil; light-colored silt loam soils have 8 to 12 meq; dark-colored silt loam soils have 15 to 22 meq; and clay soils have 18 to 30 meq.

Cation-exchange capacity facilitates retention of positively charged chemical elements from leaching, yet it gives nutrients to a growing plant root by an exchange of hydrogen (H). Farming practices that reduce soil erosion and maintain soil humus favor the maintenance of cation-exchange capacity. The cationexchange capacity of organic residues is low but increases as the residues convert to humus, which requires from 5 years to centuries.

PLANT ANALYSES

Plant analyses can be useful in diagnosing problems, in identifying hidden hunger, and in determining whether current fertility programs are adequate. For example, they often provide more reliable measures of micronutrient and secondary nutrient problems than do soil tests. When making a plant analysis to diagnose a problem, select paired samples of comparable plant parts representing the abnormal and normal plants. Abnormal plants selected should represent the first stages of a problem.

When using the technique to diagnose hidden hunger in corn, sample several of the leaves opposite and below the ear at early tassel time. For soybeans, sample the most recent fully developed leaves and petioles at early podding. For alfalfa, sample the upper 6 inches of stems and leaves at early bloom. Samples taken later will not indicate the nutritional status of the plant. After collecting the samples, deliver them immediately to the laboratory. They should be air dried if they cannot be delivered immediately or if they are going to be shipped.

Environmental factors may complicate the interpretation of plant analysis data. The more information provided concerning a particular field, the more reliable the interpretation will be. Suggested critical nutrient levels are provided in Table 2. Lower levels may indicate a nutrient deficiency.

SPAD Meter

It is sometimes called the green meter or chlorophyll meter, this

Table 2: Suggested Critical Plant Nutrient Levels for Corn and

Crop	Plant part	N	P	K	Ca	Mg	S	Zn	Fe	Mn	Cu	B
		percent						*ppm*				
Corn	Leaf opposite and below the ear at tasseling	2.9	0.25	1.90	0.40	0.15	0.15	15	25	15	5	10
Soybeans	Fully developed leaf and petiole at early podding	...	0.25	2.00	0.40	0.25	0.15	15	30	20	5	25
Alfalfa	Upper 6 inches at early bloom	...	0.25	2.00	1.00	0.25	0.22	15	25	20	7	25

N = nitrogen, P = phosphorus, K = potassium, Ca = calcium, Mg = magnesium, S = sulfur, Zn = zinc, Fe = iron, Mn = manganese, Cu = copper, B = boron.

instrument is used to measure the relative greenness of leaf tissue. Since greenness can be related to nitrogen level in the plant, using this meter, especially where nitrogen has been applied to a reference strip of the same corn hybrid at a rate somewhat above that necessary for optimal growth, can help identify deficiency in-season. This technique is effective on irrigated fields where additional nitrogen can be applied through the irrigation system with little application cost and without damage to the crop. This technique can also be useful for areas of fields that have had significant nitrogen loss due to excessive water.

Tillage Systems

Fertiliser management will be affected by tillage systems because relatively immobile materials such as limestone, phosphorus, and potassium move slowly in most soils unless they are physically mixed by tillage operations. Such "stratification" of nutrients, with higher concentrations developing near the surface, has been well documented in a number of studies but has not been shown to reduce yields of corn or soybeans in Illinois. Limited research indicates that plants develop more roots near the soil surface in conservation-tillage systems, due apparently to both the improved moisture conditions caused by the surface mulch of crop residues and the higher levels of available nutrients.

With continued reduced-tillage practices, soil fertility levels at deeper depths may be depleted such that future soil fertility practices may need adaptation. Soil tests are important for phosphorus, potassium, and limestone management under any tillage system. Nitrogen fertiliser management may be affected to a limited extent by changing tillage systems.

Starter Fertiliser

Starter fertiliser is more effective than broadcast applications under cool, moist conditions when phosphorus soil test levels are low, irrespective of tillage system. At high soil test levels, starter fertiliser often results in early growth response on conventional tillage systems but seldom results in increased yield at harvest. Early season growth of no-till corn is frequently less vigorous than conventional tillage. This slower growth is likely the result of cooler soil temperatures and higher soil moisture conditions associated with the high residue mulch. Both of these conditions tend to slow root growth and thus the ability of the plant to absorb nutrients.

Lime

Soil acidity is one of the most serious limitations to crop production. Acidity is created by removal of bases by harvested crops, leaching, and an acid residual that is left in the soil from nitrogen fertilisers. During the last several years, limestone use has tended to decrease in Illinois while crop yields and nitrogen fertiliser use have increased. At the present rate of limestone use, no lime is being added to correct the acidity created by the removal of bases or the acidity created in prior years that has not been corrected. A soil test every 4 years is the best way to check on soil acidity levels.

Soil Acidity

Soil acidity affects plant growth in several ways. Whenever soil pH is low (and acidity is high), several situations may exist:

— The concentration of soluble metals may be toxic. Damage from excess solubility of aluminum and manganese due to soil acidity has been shown in field research.

— Populations and the activity of the organisms responsible for transformations involving nitrogen, sulfur, and phosphorus may be altered.

— Calcium may be deficient. This usually occurs only when the cation-exchange capacity of the soil is extremely low.

— Symbiotic nitrogen fixation in legume crops is impaired greatly. The symbiotic relationship requires a narrower range of soil reaction than does the growth of plants not relying on nitrogen fixation.

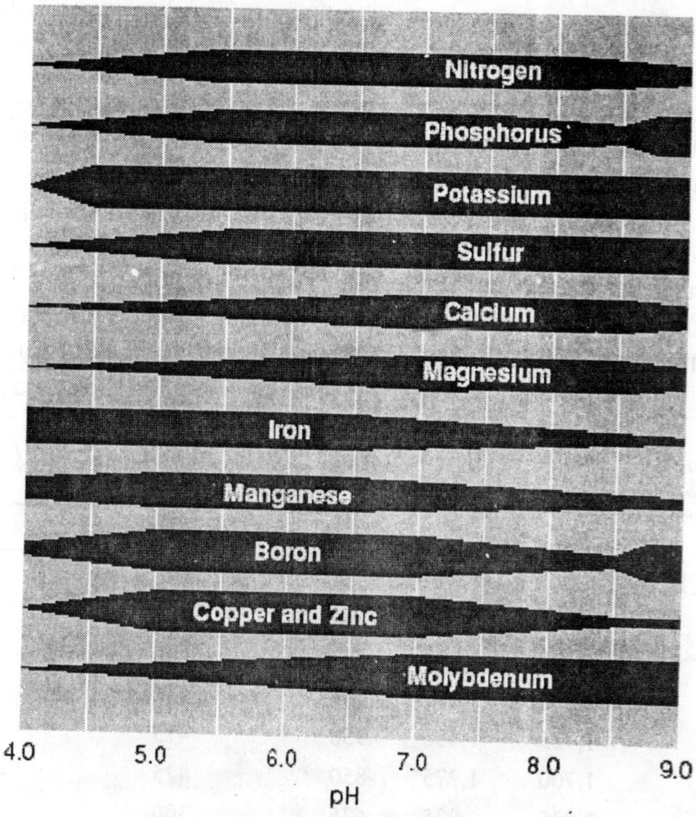

Figure 1: Available nutrients in relation to pH.

— Acidic soils are poorly aggregated and have poor tilth. This is particularly true for soils that are low in organic matter.

— The availability of mineral elements to plants may be affected.

Figure 1 shows the relationship between soil pH and nutrient availability. The wider the dark bar, the greater the nutrient availability. For example, the availability of phosphorus is greatest in the pH range between 5.5 and 7.5, dropping off below 5.5.

Because the availability of molybdenum is increased greatly as soil acidity is decreased, molybdenum deficiencies usually can be corrected by liming.

While high-pH soils (pH > 7.4) result in reduced availability of several nutrients, particularly phosphorus, zinc, iron, and manganese, decreasing soil pH has not been shown to be economical for the production of agronomic crops. Acidification of soils to produce crops such as blueberries and cranberries is essential if the pH is high. Acidification can be accomplished by applying elemental sulfur, aluminum sulfate, or iron sulfate. The amount of elemental sulfur needed to reduce soil pH depends on the initial pH and the desired pH (Table 3).

Table 3: Amount of Elemental Sulfur Needed to Reduce Soil pH Soil group1 Original

pH	A	B	C	D
Elemental sulfur (lb/A) needed to reach pH 5.0				
6.4	2,700	2,100	1,400	700
6.2	2,400	1,800	1,200	600
6.0	2,150	1,625	1,075	550
5.8	1,925	1,450	950	475
5.6	1,700	1,275	850	425
5.4	1,225	925	625	300
5.2	775	575	375	200
5.0				
Elemental sulfur (lb/A) needed to reach pH 4.5				
6.4	4,000	3,000	2,000	1,000
6.2	3,800	2,800	1,900	950
6.0	3,525	2,650	1,775	925
5.8	3,300	2,475	1,650	825

5.6	3,075	2,300	1,525	775
5.4	2,600	1,950	1,300	650
5.2	2,150	1,625	1,075	550
5.0	1,375	1,050	700	350

Method of Fertiliser Application

With the advent of new equipment, producers have a number of options for placement of fertiliser. These options range from traditional broadcast application to injection of the materials at varying depths in the soil. Selecting the proper application technique for a particular field depends at least in part upon the inherent fertility level, the crop to be grown, the land tenure, and the tillage system. On fields where the fertility level is at or above the desired goal, there is little research evidence to show any significant difference in yield that is associated with method of application. In contrast, on low-testing soils and in soils that "fix" phosphorus, placement of the fertiliser within a concentrated band has been shown to result in higher yields, particularly at low rates of application. On higher-testing soils, plant recovery of applied fertiliser in the year of application is usually greater from a band than a broadcast application, though yield differences are unlikely.

Broadcast fertilisation: On highly fertile soils, both maintenance and buildup phosphorus and potassium are efficiently utilized when broadcast and then plowed or disked in. This system, particularly when the tillage system includes a moldboard plow every few years, distributes nutrients uniformly throughout the entire plow depth. As a result, roots growing within that zone have access to high levels of fertility. Because the nutrients are intimately mixed with a large volume of soil, opportunity exists for increased nutrient fixation on soils having a high fixation ability. Fortunately, most Illinois soils do not have high fixation rates for phosphorus or potassium.

Row fertilisation: On soils of low fertility, placement of fertiliser in a concentrated band below and to the side of the seed has been shown to be an efficient method of application, especially in situations for which the rate of application is markedly less than that needed to build the soil to the desired level. Producers who are not assured of having long-term tenure on the land may wish to consider this option. The major disadvantages of this technique are (1) the additional time and labor required at planting time; (2) limited contact between roots and fertiliser; and (3) inadequate rate of application to increase soil levels for future crops.

Strip application: With this technique, phosphorus, potassium, or both are applied in narrow bands on approximately 30-inch centers on the soil surface, in the same direction as the primary tillage. The theory behind this technique is that, after moldboard plowing, the fertiliser will be distributed in a narrow vertical band throughout the plow zone. This system reduces the amount of soil-to-fertiliser contact as compared with a broadcast application, and thus it reduces the potential for nutrient fixation. Because the fertiliser is distributed through a larger soil volume than with a band application, the opportunity for root-fertiliser contact is greater.

Deep fertiliser placement: Several terms have been used to define this technique, including root-zone banding, dual placement, knife injection, and deep placement. With this system, a mixture of nitrogenphosphorus or nitrogen-phosphorus-potassium is injected at a depth from 4 to 8 inches. The knife spacings may vary by crop to be grown, but generally they are 15 to 18 inches apart for close-grown crops such as wheat and 30 inches for row crops.

Dribble fertiliser: This technique applies urea-ammonium nitrate solutions in concentrated bands on 30-inch spacings on the soil surface. Results from several states have shown that this system reduces the potential for nitrogen loss of these materials, as compared with an unincorporated broadcast application. However, it has not been shown to be superior to an injected or an incorporated application of urea- ammonium nitrate solution.

"Pop-up" fertilisation: The term "pop-up" is a misnomer. The corn does not emerge sooner with this kind of application, and it may come up 1 or 2 days later. The corn may, however, grow more rapidly during the first 1 to 2 weeks after emergence. Pop-up fertiliser will make corn look very good early in the season and may aid in early cultivation for weed control. But no substantial difference in yield is likely in most years due to a pop-up application as compared to fertiliser that is placed in a band to the side and below the seed. Seldom will there be a difference of more than a few days in the time the root system intercepts fertiliser placed with the seed as compared to that placed below and to the side of the seed.

Under normal moisture conditions, the maximum safe amount of N plus K_2O for pop-up placement is about 10 or 12 pounds per acre in 40-inch rows and correspondingly more in 30- and 20-inch rows. In excessively dry springs, even these low rates may result in damage to seedlings, reduction in germination, or both. Pop-up fertiliser is unsafe for soybeans. In research conducted at Dixon Springs, a stand was reduced to one-half by applying 50 pounds of 7-28-14 and reduced to one-fifth with 100 pounds of 7-28-14. Site-specific application.

Equipment has recently been developed that uses computer technology to alter the rate of fertiliser application as the truck passes across the field. This approach offers the potential to improve yield while minimizing the possibility of overfertilisation. Yield improvement results from applying the correct rate (not a rate based on average soil test) to the low-testing portions of the field. Overfertilisation is reduced by applying the correct rate (in many cases zero) to high-testing areas of the field. The combination of improved yield and reduced output results in improved profit.

Foliar fertilisation: Researchers have known for many years that plant leaves absorb and utilize nutrients sprayed on them.

Foliar fertilisation has been used successfully for certain crops and nutrients. This method of application has had the greatest use with nutrients required in only small amounts by plants. Nutrients required in large amounts, such as nitrogen, phosphorus, and potassium, have usually been applied to the soil rather than the foliage.

POTASSIUM

Potassium is retained by soils more strongly than N, it can be lost through leaching, and may need replacing. Prompt replacement is important, especially early in the season. Approximately 25 to 30 pounds per acre of potash should correct most leaching losses. The symptoms of potassium deficiency can be very pronounced. Deficiency symptoms first appear on the older leaves as a yellowish-white mottling. The mottling changes to a light yellowish green, and yellow spots appear between veins. The centers of these spots die, and numerous brown specks appear at the leaf top, around the margin, and between the veins. The tip and the margin of the leaf breakdown first and curl downward. As this physiological breakdown progresses, the whole leaf becomes reddish brown, dies, and is shed prematurely. The premature shedding of leaves contributes to dwarfed and immature bolls.

In recent years, some regions have reported K deficiency symptoms appearing in the upper part of the plant. In some cases, soil K levels appear to be high, but the plants are unable to obtain adequate K. In these cases, foliar K fertilisation has improved yield and quality. At the present time, these symptoms have been associated with three factors:

1) the use of very high yielding, determinate-type cultivars that set a heavy fruit load over a very short period;
2) soils that "fix" K in non-available forms, and
3) an unidentified disease.

POTASSIUM DEFICIENCY IN COTTON LINKED TO LEAFSPOT DISEASE

In almost every case where leafspot was investigated, low soil K, low plant tissue K and/or low petiole K was discovered. Low petiole K appeared to be the best indicator. In some cases the problem was traced back to inadequate K fertilisation. Some cases were under dryland conditions where low soil moisture was suspected of reducing uptake of K. Some cases occurred on short-season varieties where intense demand for K in a short period of time was suspected as the main problem. A few cases occurred with high soil magnesium (Mg) levels which were thought to have caused competition for K uptake and subsequent K deficiency.

The majority of cases, however, were discovered on full season varieties under irrigation around the fourth week of bloom with heavy fruit set. This timing and situation correspond with a heavy demand for K. The roots of the cotton plant also start to decline at this time due to competition for carbohydrates by developing bolls. This adds to the challenge of taking up soil K at this time. Even with irrigation, adequate water may not have been provided during a critical dry period, or with adequate water, may have contributed to higher yield conditions and K demand. Once K deficiency sets in and leafspot appears, fungicide sprays do not alleviate the condition since the primary problem is K deficiency.

If K deficiency is detected around the fourth week of bloom and is not severe, foliar K sprays may lessen yield effects. Petiole testing could also help avoid this problem, since it is designed to predict nutrient deficiencies up to two weeks in advance, especially as the crop moves toward peak bloom. Unfortunately, if severe K deficiency occurs late (sixth week of bloom) foliar K sprays will likely not correct the problem. Also, K deficiency and leafspot are fairly common once cotton "cuts-out". No corrective treatment is recommended at this time.

BEST MANAGEMENT PRACTICES TO AVOID POTASSIUM DEFICIENCY

— *Soil Testing* : The first and best line of defense for avoiding K-deficient cotton is soil testing. Maintaining soil test K levels in the medium to high range for cotton is recommended. Also keeping a good balance of other nutrients such as calcium (Ca) and Mg will help.

— *Split K Applications* : Since K is relatively mobile in sandy soils, split applications are recommended on soils with no clay subsoil in the top 16 inches. Apply half the K at planting and remainder at side-dressing, sometime around the first square. This helps supply K at a time when demand increases rapidly and may even be helpful on "stiffer" soils.

— *Foliar Fertilisation and Petiole Testing* : In most cases where soil K levels are maintained at medium to high levels, preplant soil applications of K fertiliser should provide enough K so that foliar applications will not be necessary. There are a number of cases, however, where a yield response to foliar applications may occur: deep sands, low-soil K at planting, high yield irrigated conditions, and during periods of limited soil moisture. The best way to determine the need for foliar K is by petiole testing. A complete petiole testing program is designed to predict nutrient deficiencies up to two weeks in advance, before any yield reductions due to deficiencies occur.

Excessive fertiliser rates should not be used as a strategy to avoid K deficiency. Other problems can result from nutrient imbalance. Getting back to the basics of soil testing, proper fertilisation, and petiole testing should help eliminate K deficiency as a cause of yield reduction for cotton in the future.

Chapter 11

Fertilisation and Disease

Mineral elements are, in fact, directly involved in all mechanisms of defense as integral components of cells, substrates, enzymes and electron carriers or as activators, inhibitors and regulators of metabolism. Tolerant or moderately resistant plants demonstrate the greatest response to mineral elements, while disease reactions of highly resistant or highly susceptible plants are not as readily altered by nutrition. The availability of mineral elements to trees and their effect on disease depends on their form and solubility, on the presence of competing or toxic entities, and on environmental factors such as pH, moisture, temperature and aeration.

Researchers studied the effect of N ($(NH_4)_2NO_3$) fertilisation on Cytospora canker in aspen in a greenhouse hydroponic system. Aspen trees are commonly used in western U.S. landscapes, but they are susceptible to infection by several canker-inducing pathogens, including Cytospora chrysosperma. Cytospora canker is also common on many willows and other poplars. Although this disease is stable in native or naturalized areas, canker incidence appears to be increasing in maintained urban landscapes. Conditions that stress host trees influence incidence and expansion of cankers caused by C. chrysosperma. These stresses include drought, transplant stress, pruning wounds, insect damage, excess soil salts and severe defoliation.

In this research, researchers grew the aspens in sand and fertilised them with 1 of 5 N treatments for 2 growing seasons. Nitrogen deficiency contributed to significantly larger cankers. Among trees receiving moderate levels of nitrogen cankers failed to expand which suggests that tree resistance mechanisms may involve host response to nutrient deficiencies, rather than fungal stimulation via nutrition. Cankers expanded at high levels of N indicating that excess N may stress aspen. But, because this rate is not normal, researchers point out that excess N is not likely a predisposing stress of aspen.

The addition of nitrogen encouraged pathogenesis in an experiment conducted by Entry et al in which second growth stands Douglas firs were thinned and fertilised with 360 kg of N (as urea) per hectare. Ten years later after treatment, trees were inoculated with two isolates of Armillaria ostoyae. Results demonstrated that this treatment predisposed the Douglas firs to infection by A. ostoyae by lowering concentrations of defensive compounds in root bark and increasing the energy available to the fungus to degrade them. They hypothesized that trees growing extremely fast may allocate more carbon to sugar and cellulose and less carbon to tree defense compounds, such as lignin, phenolics and tannins. Citing research state that fungi can degrade phenolic compounds only when an additional carbon source is present; the rate of degradation is directly proportional to the amount of additional growth substrate. Carbon utilization by A. ostoyae in culture was more efficient at low sugar concentrations, but fungal biomass was greater at higher sugar concentrations.

INSECTS

Tree damaging insects can also be affected by fertilisation. Researchers fertilised eastern hemlock with nitrogen in a forest plantation in Connecticut and found that it stimulated population growth of the hemlock wooly adelgid. Fertilised hemlocks had five times more adelgids, had inferior color, and produced 25%

less new growth than unfertilised trees after a single adelgid generation. These trends did not differ between hemlocks which had been fertilised 6 months prior to infestation by A. tsugae and those which were fertilised at the same time that trees were infested. They concluded that N fertilisation of hemlock neither increased host resistance to the adelgid nor repressed adelgid population growth following establishment. These results may be generally applicable to piercing and sucking insects that feed on trees and shrubs.

Chlorosis

Chlorosis is characterized by yellow leaves; leaves may become progressively smaller as a result of shortage of chlorophyll and food production in the leaf. As the condition worsens necrotic areas may be observed between the veins and shoot growth may be stunted. Commonly, this condition is attributed to a deficiency of either iron or manganese caused by high soil pH but may be caused by reduced availability of one or more of the soil nutrient elements such as N, K, Mg, B, Zn, Cu and Md Even excessive amounts of some elements may cause chlorosis. Other factors such as low temperatures, reduced sunlight, high soil moisture, and excessive applications of calcium and possibly phosphorous in fertilisers and irrigation water can also cause the development of chlorosis. Historically, certain tree species have exhibited habitual chlorosis, particularly when planted along streets and around homes where the original topsoil has been removed or mixed with the subsoil.

Oaks and maples are by far the most susceptible, with oaks heading the list, particularly pin oak. Many studies have evaluated treatments for correcting chlorosis and have found iron injections and implants to be very effective. Some investigators have researched methods of acidifying the soil to correct chlorosis induced by unavailable iron and manganese in soils above pH 6.2. Nutrient imbalances and normal leaf color can be restored and maintained for several years by topsoil and subsoil treatment with sulfuric acid. Sulfuric acid diluted in 5 gallons of water/ 1000 sq. ft. in 2 inch diameter holes, 2 feet apart, in two circles beneath the crown. The pH of topsoil beneath treated pin oaks was approximately neutral 3 years after application of the sulfuric acid. Subsoils receiving similar treatments were still quite acidic after 4 years.

Maple Dieback

It is also known as maple decline and maple blight. This is an another disease of urban trees, which is more a description of symptoms than a specific malady. These symptoms include chlorotic and scorched leaves that are often smaller than normal, premature fall coloration and leaf drop, and branch dieback initially involving the upper crown . Among the documented causes are road salts, nitrogen deficiency, high pH-manganese deficiency complex and drought. Root rots and cankers further contribute to the decline. Research on sugar maples showed that leaf color can be significantly improved by high nitrogen fertilisation.

Nutrient level was lower in chlorotic leaves than in healthy leaves for all of the elements listed except sodium and aluminum. The extremely high sodium level found in chlorotic leaves, they concluded, may implicate salt in maple decline along streets and highways. Prolonged nutrient deficiency stresses urban trees and may eventually lead to disease and increased susceptibility to insect attack. Addition of nutrients can bolster resistance against disease expression and can increase tree vigor. Excessive or unneeded application of nutrients, however, may increase pathogenesis and/or increase insect feeding on succulent tissue.

NITROGEN ADDITION TECHNIQUE

Nitrogen Addition Technique is a new approache to Fertilisation Theory. The study of plant nutrition has historically focused on the nutrient concentration in solution as the general driving variable for nutrition of plants; thus creating the impression that the rate of ion uptake depends on the external ion concentration. Some researchers using a nitrogen addition technique adjusted to the rate of plant growth and consumption demonstrated that high as well as low growth rates of birch seedlings may be obtained at the same very low N concentration in the solution.

When optimum nitrogen supplies were reduced to little or none, typical nitrogen deficiency symptoms resulted. However, if nitrogen was then supplied at a rate proportional to the increase in plant growth, even though nutrient levels were below optimum, growth stabilized at a slower rate, leaf color returned, and the root shoot ratio stabilized higher than when the seedlings were growing under more optimum conditions. In addition, the photosynthetic rate and the amount of chlorophyll were not reduced in proportion to the reduction in the nitrogen supply or the overall growth of the plant. The nutrient concentration concept that has been the criterion for fertiliser application today has its consequences. In addition to negative effects on the environment through leaching, the high salt concentrations resulting from one or only a few nutrient applications during the growth season may result in root injury. The nitrogen addition technique may offer the possibility to achieve high fertiliser efficiency at the same time as the nitrogen requirements throughout the whole period of growth are satisfied.

NUTRIENT REQUIREMENTS DETERMINING

To estimate how well the nutritional requirements of a tree are being met four methods of analysis are generally employed: visually observing the tree's growth and appearance, testing the soil, or testing the foliage (or other plant tissues), and various combinations of the three.

Visual Observation of the Tree's Growth and Appearance

A severe deficiency of any essential element is usually

accompanied by symptoms which may be detected visually; however, visual nutrient deficiency symptoms can be variable, complex and often not easily distinguishable from one another, from air pollution or other tree stress symptoms. Most authorities, therefore, recommend foliar, soil or other forms of analyses to assist in or substantiate visual diagnosis.

Nutrient Deficiency Symptoms: There is voluminous literature dealing with nutrient deficiency symptoms of plants and trees. Table 1 is a compilation of the most frequently cited and definitive symptoms associated with essential nutrient deficiencies.

Table 1: Nutrient Deficiency Symptoms for Trees

Nitrogen (N)

Broadleaf: smaller, thinner, fewer leaves, general yellowish green color, more pronounced on older leaves; poor or stunted shoot growth; early leaf drop, thin crowns

Conifer: yellow, short needles that are close together; older plants exhibit poor needle retention; lower crowns may be yellow while upper crowns remain green

Phosphorus (P)

Broadleaf: dark green, blue green, slightly smaller leaves. Veins, petioles, or lower surface may become reddish-purple, foliage is less dense; stunting or poor growth occurs prior to reddening of leaves, older leaves show symptoms first and most severely.

Conifer: needles turn purple in young seedlings; needles of spruce and larch turn gray or bluish-green; symptoms most pronounced in later summer on older needle tips; roots are sparse with no mycorrhizae.

Potassium (K)

Broadleaf: partial chlorosis of most recently matured leaves in interveinal area beginning at tips, followed by necrosis; older leaves may become brown and roll upward; often irregular necrotic spots or lesions on leaves; may exhibit dark bronze leaf colors.

Conifer: small, yellow-green needles; most pronounced on tips of older needles in autumn, winter and spring; needle retention is poor; seedlings have short, thick, abundant buds; frost injury is frequent.

Calcium (Ca)

Broadleaf: death of terminal buds, tip dieback; leaves chlorotic and/or necrotic; leaves may be brittle and stiff; young leaves distorted and small.

Conifer: primary needles are usually normal, but secondary needles may be stunted or killed.

Magnesium (Mg)

Broadleaf: marginal chlorosis on older leaves followed by interveinal chlorosis; shoot growth not affected until deficiency is severe; symptoms disappear quickly after fertilisation.

Conifer: golden or yellow-tip "halo" effects on conifer needles in late summer or autumn; sharp transition to the green portion; symptoms more severe in moist years.

Sulfur (S)

Broadleaf and Conifer: symptoms similar to those of N deficiency; yellow-green or yellowish foliage, especially in younger leaves; reduced shoot growth; older leaves usually not affected.

Iron (Fe)

Broadleaf: Interveinal chlorosis of young leaves, especially in wet or cool years. In oaks, young leaves may be yellow on emergence; develop interveinal necrotic spots and light color; and finally curl, wither and die. Exposed leaves bleached.

Conifer: new growth will be very stunted and chlorotic; older needles and the lower crown will remain green.

Manganese (Mn)

Broadleaf: marginal leaf chlorosis, gradually extending between the major veins, with bands of green along the main veins and the midrib; necrotic spots may develop in the chlorotic areas; shoot growth may be reduced.

Conifer: symptoms essentially the same as for iron deficiency; new needles are chlorotic and pale green; tip necrosis may occur.

Zinc (Zn)

Broadleaf: marked chlorosis of younger leaves; may be uniformly yellow, sometimes mottled with necrotic spots; leaves are small ("little leaf") and may be deformed; shoot dieback in severe cases; may be rosettes of leaves at the shoot tip.

Conifer: extreme shortening of branches, needles, and needle-spacing may occur in upper crown, plus general yellowing and loss of older needles; terminals die back.

Boron (B)

Broadleaf and Conifer: death or distortion of meristematic tissues; terminal growth dies; may be tip wilting, bending, shoots may be short, brushy, stiff; young leaves may be red, bronzed or scorched in broadleaf species.

Copper (Cu)

Broadleaf: most common symptom is stunting of over-all growth followed by leaf stunting, loss of leaf luster and leaf size; rosetting of buds on terminal branches, terminal growth may die

Conifer: young pine needles show tip burn; shoots of Douglas fir are week and often crooked; needles at the tips of shoots may discolor and drop during winter.

Molybdenum (Mo)

Broadleaf: cupping of the older leaves; marginal chlorosis followed by interveinal

chlorosis; leaves are similar in color to those deficient in N; shoot internodes are stunted when deficiency is severe.

Analysis of the Soil

Analysis of the soil gives information about the availability of essential elements, the cation-exchange-capacity of the soil, organic matter levels, and soil pH. Soil analysis can also aid in evaluating nutrient status and need for fertilisation. Soil analysis information on nutrient levels is usually expressed in parts per million or pounds per acre and is less expensive than foliar analysis Soil analysis, although helpful, is not always a good index of the nutritional status of a tree. Soil analysis results often vary significantly across a given year for certain elements. The potential lack of consistent and accurate data make it difficult to calculate amounts of nutrients to add.

In addition, the efficiency with which a tree utilizes the soil nutrients varies widely with different soil conditions. Researcher demonstrated this in research and conducted on apple trees in New York in which some soils with rather low amounts of K produced trees with high K content and vice versa. That is, even though we can calculate the amount of fertiliser to increase the soil nutrient content, no assurance that this will increase our tree nutrient content. Finally, it is important to note that extractants are used to determine the availability of specific nutrients. That the flushing of extractants through soil duplicates the roots ability to take up these same nutrients under the same conditions is a large assumption to make. In addition, the use of extractants is based on agricultural models modified for forestry usage. Often the results of soil analyses are more difficult to interpret than the results of a foliar analysis. Some researchers support this statement by the following observations:

1) the availability of some elements depends on the physical conditions of the soil. Frequent water logging, or insufficient levels of soil oxygen, for example, have negative effects on the uptake of K and the rate of N mineralization;
2) the uptake of some nutrients, primarily N, depend on the extension of the root system; and
3) the criteria for the interpretation of the results of soil analysis have been derived from forest research - where many of the critical values for nutrient deficiencies focus on the assessment of levels at which trees react to fertilisation with an increase of growth.

These levels are generally higher than those required for healthy or "acceptable" growth and development. Therefore, soil analysis results of urban trees - for which acceptable growth and amenity value are normally more important than optimal production - are potentially less helpful than results of foliar analysis in providing an accurate explanation for tree health problems related to nutrient deficiencies.

Foliar Analysis

The purpose of foliar analysis is to evaluate nutrient status and make possible a diagnosis of nutrient deficiency or toxicity and/ or a prediction of fertilisation response. Some arborists recommend sampling mature leaves that have been exposed to full sun, collected between August and mid-September when nutrient concentration levels are more or less constant. In many areas leaves are usually analyzed for the content of the macronutrients, N, P, K, Ca, Mg and sometimes Cl. In regions where micronutrient deficiencies have not been observed, the content of these nutrients may not be analyzed.

The rate of growth and size of the plant are determined by the nutrient concentrations in the foliage, other tissues than foliage might be more diagnostically suitable, at least for certain nutrients and certain tree species. White and Leaf , in the work with pine, showed strong correlation among K contents of the total tree tissue, i.e. wood, live branches, bark, with tree height growth, but no statistically significant relations among foliage K levels and soil K levels and tree height growth. To determine the effect of N fertiliser on leaf N concentrations in valley oak and Chinese pistache and found no correlation between visual N deficiency symptoms and actual leaf N concentrations.

It is important to realize that none of the nutrient analysis methods suffices on its own and many authorities recommend that soil and foliar analyses are done together. Foliar analysis may be a more accurate way to assess deficiencies in the uptake of an element, but this method may not indicate how the deficiency can be corrected. To know whether fertilisation is of any use, more should be known about the chemical soil fertility and physical soil factors that might influence the uptake of minerals.

DIAGNOSIS AND RECOMMENDATION INTEGRATED SYSTEM (DRIS)

DRIS that incorporates relative balance among nutrients as well as nutrient concentration in foliar nutrient diagnoses. In short, DRIS provides indices that measure the relative deficiency or sufficiency of nutrients relative to each other based on their deviation from optimum levels that have been established for a desired group of trees. DRIS has successfully been used on many crops including corn, soybeans, wheat and on hybrid poplar and Monterey pine. DRIS has great potential as a nutrient diagnostic system for trees in the urban forest. Desirable characteristics of urban trees such as healthy leaf color and size, certain diameter and crown size, and general appearance could be used to divide and rank trees into grades or groups.

Following DRIS methodology, comparisons of nutrient ratio means and variances could then be made among the groups and important information regarding nutrient quantity and balance could be obtained. This information could then be effectively incorporated into a judiciously applied fertilisation program if nutrients were found lacking. Some researchers used DRIS in a study of a Fraser fir Christmas tree plantation in North Carolina. They discovered that in contrast to a critical-level approach where higher nutrient levels would be associated with either decreased deficiency or increased sufficiency, lower foliar Mg levels corresponded with improved Christmas tree performance. In fact, three out of the five nutrients examined (P, Ca and Mg) had lower mean concentrations in premium versus nonpremium trees.

Tree quality is a function of nutrient balance or synchronization as well as absolute levels of individual nutrients.

What appears to be the result of a deficiency of one nutrient may as well be the result of an excess of another. Element relationships should be considered when contemplating any addition of nutrients to soil (Table 2). For example, the ratio of calcium to magnesium available for plant growth is important. If calcium is in excess, magnesium absorption is depressed or vice versa. High levels of P can induce micronutrient deficiencies, such as copper, iron and zinc. Also, the effect of large quantities of K decreasing Mg is widely recognized. Kuhns reports that when too much K causes nutrient imbalance, plant leaves are large, but relatively inefficient at photosynthesis.

The resulting abnormally large concentration of N compounds compared to photosynthates (carbohydrates) in the leaves makes the leaves susceptible to fungal and bacterial diseases and drought stress. Sometimes it is the order in which a nutrient is added that affects the uptake and availability of other nutrients. The application sequence of nitrogen greatly influenced the availability of soil and fertiliser K. When NH_4^+ was applied alone or after K it reduced K uptake, but if applied prior to K, it enhanced K uptake.

Table 2: Important Relationships of Nutrient Elements.

An excess of....	may cause a deficiency of....
nitrogen	potassium
phosphorus	iron, copper, zinc, and manganese
potassium	calcium and/or magnesium
sodium	potassium, calcium, and magnesium
calcium	magnesium, boron, iron, manganese and copper
magnesium	calcium
iron	manganese, molybdenum
manganese	iron

Among the most commonly observed antagonisms is that of the effect of N fertilisers in apparently decreasing content of K and P and sometimes increasing Mg. The majority of these results, and other similar reported results for other nutrients, are based on leaf analysis and the assumption that any change in leaf concentration reflects like changes in total plant content. Researchers experimented with the effect of N on the distribution of P, K and Mg in New York apple trees and found that K and P are reduced in the leaves and that Mg is increased as a result of increased nitrogen fertiliser. They studied various plant tissues and found that with additional nitrogen the relative changes in the distribution of minerals and dry weight in various tissues resulted in either an increase or decrease in percentage composition with an overall increase in total absorption.

Mg distribution response did vary between plants with high K and those with low K during the growing season - K deficient trees transported Mg more rapidly to the leaves and constantly depleted their roots of the nutrient. However, at the end of the growing period the roots of both sets of trees had approximately the same quantity of Mg but the leaves of the trees grown with low K had nearly twice as much. These data shed a little light on the antagonism picture of N, K and Mg.

They show not only that these nutrient elements are unevenly distributed throughout the plant but that changes resulting in one plant part do not necessarily reflect corresponding changes in another, or the entire plant. It also suggests that fertilising trees without knowing the nutritional status is not only costly but may well have deleterious effects.

NUTRIENT UPTAKE

Tree nutrient absorption begins in the soil where roots absorb nutrients as charged particles or ions. Regardless of the type of soil or location of the absorbing root, certain aspects of the nutrient absorption process are common to all tree species, although variations in degree can occur. Nutrient ions are transported to and into the roots in one of three ways:

1) they may be delivered in the bulk movement of the soil solution as the result of transpiration;

2) they may diffuse along a gradient of high nutrient concentration to one of a lower concentration; and

3) the root may arbitrarily grow to the nutrient source.

The significance of each process varies with the nutrient under consideration but more than likely all three processes are involved in nutrient absorption by trees. Bulk flow is dependent upon the amount of soil moisture transpired by the tree and on the concentration of nutrients in solution. Diffusion takes place because absorption during periods of low transpiration lowers the concentration at the root surface and creates a gradient along which the nutrients move from the soil solution toward the roots.

Plant roots, also, absorb and accumulate nutrient ions selectively - often against a concentration or electrical gradient. The metabolic expenditure for some nutrient uptake can be large. The roots deprived of nitrogen for 10 days respired at only 71% of the rate for roots supplied with nitrogen. The outermost zone of the root surface is described as a negatively charged mucilaginous layer. This layer is easily visible on young roots of many trees, especially pines. Passive uptake includes movement of nutrients through this layer, through parts of the cell walls, as well as through a portion of the cytoplasm.

In the region between the ectoplasm and the tonoplast of the cell, the active energyrequiring process of ion uptake dominates and both cations and anions are thought to be transported across the tonoplast into the cell vacuole by a complex carrier system. It is not yet known whether each specific ion has its own carrier, but it seems that this is likely - at least for nutrients absorbed in large amounts, like nitrate, phosphate and potassium. It is possible that there are dual absorption systems for many ions and that ions with similar properties must compete for the same absorption sites.

Absorption rates of specific ions are different for different species and often within species there can be wide genetic differences in the ability to accumulate nutrients. Once inside, the cell may use the nutrient ion for its own metabolic needs or deliver the ion out into the xylem. Driven by transpiration, cells along the way may remove nutrients from the xylem flow as needed, using the same processes that roots used to absorb them.

BIBLIOGRAPHY

Appl. Genet. 106: 568-574. Lu, C.S., and Bridgen, M.P., Chromosome doubling and fertility study of *Alstroemeria aurea x A. caryophyllaea*. Euphytica 94 1: 75-81, 1997.

Darwin, C. *The Various Contrivances by which Orchids are Fertilised by Insects*. London, J. Murray, 1904.

______________, *The Variation of Animals and Plants under Domestication*. London, John Murray, 1905.

______________, *Insectivorous Plants*. London, J. Murray., 1876.

______________, *The Effects of Cross and Self Fertilisation in the Vegetable Kingdom*. London, J. Murray, 1876.

______________, *The Power of Movement in Plants*. London, Murray, 1880.

______________, *The Movements and Habits of Climbing Plants*. London, Murray, 1882.

Martin, F.W., Compounds of the stigmatic surface of *Zea mays* L. Ann. Bot. 34: 835–842, 1970.

Matzk, F., Hammer, K. and Schubert I., "Coevolution of apomixis and genome size within the genus *Hypericum*". *Sexual Plant Reproduction* 10.1007/s00497-003-0174-8, 2003.

Merlin, C.M. and Grant,W.F., *Hybridisation studies in the genus Impatiens*. Can. J. Bot. 64: 1069–1074., 1986.

Mok, D.W.S. and Peloquin, S.J., *The inheritance of three mechanisms of diplandroid (2n-pollen) formation in diploid potatoes*. Heredity 35: 295–302, 1975.

Morgan, E.R., Burge, G.K., Seelye, J.F., Hopping, M.E., Grant, J.E., Warren, A.G.F. and Brundell, D., Wide crosses in the Colchicaceae: *Sandersonia aurantiaca* (Hook.) x *Littonia modesta* (Hook.) Euphytica 121: 343-348, 2001.

Ohri D. and Khoshoo T.N., "Cytogenetics of garden gladiolus, IV", *Origin and evolution of ornamental taxa*. Proc. Indian Natn. Sci. Acad. B49 No 3: 279-294, 1983.

Okada, H., Polyphyletic allopolyploid origin of *Ranunculus cantoniensis* (4x) from *Ranunculus silerifolius* (2x) × *Ranunculus chinensis* (2x). Plant Syst. Evol. 148: 89-102, 1984.

Ramanna, M.S., *The role of sexual polyploidisation in the origins of horticultural crops: Alstroemeria as an example*. In: A. Mariani and S. Tavoletti (eds.), Proceedings of Workshop: Gametes with Somatic Chromosome Number in the Evolution and Breeding of Polyploid Polysomic Species: Achievements and Perspectives, Tipolitografia Porziuncola–Assisi (PG) Italy, pp 83-89, 1992.

Sharma, D.R., Kaur, R. and Kumar, K., *Embryo rescue in plants - A review*. Euphytica 89: 325-337, 1996.

Van Creij, M.G.M., Kerckhoffs, D.M.F.J. and Van Tuyl, J.M., *The effect of ovule age on ovary-slice culture and ovule culture in intraspecific and interspecific crosses with Tulipa gesneriana* L. Euphytica 108: 21-28, 1999.

Van Eijk, J.P., Van Raamsdonk, L.W.D., Eikelboom, W. and Bino, R.J., "Interspecific crosses between Tulipa gesneriana cultivars and wild Tulipa species - A Survey". *Sex. Plant Reprod.* 4: 1-5, 1991.

INDEX